SO-ABB-753

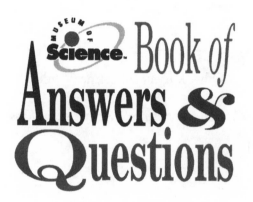

Other Museum of Science™ titles:

Museum of Science™ Activities for Kids

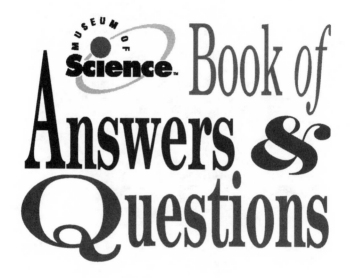

Museum of Science Book of Answers & Questions

Ann Rae Jonas

ADAMS MEDIA CORPORATION
Holbrook, Massachusetts

Copyright ©1996, Museum of Science. All rights reserved.
This book, or parts thereof, may not be reproduced in any form
without permission from the publisher; exceptions are made for brief excerpts
used in published reviews.

Published by Adams Media Corporation
260 Center Street, Holbrook, MA 02343

ISBN: 1-55850-645-4

Printed in the United States of America.

First Edition
J I H G F E D C B

Library of Congress Cataloging-in-Publication Data
Jonas, Ann Rae.
Museum of science book of answers & questions / developed by The Museum of Science
in Boston ; written by Ann Rae Jonas.
p.   cm.
Includes bibliographical references and index.
ISBN 1-55850-645-4
1. Science. 2. Science—Miscellanea. I. Boston Museum of Science. II. Title.
Q158.5.J66   1996
500—dc20   96-2888
CIP

This publication is designed to provide accurate and authoritative information with regard to the
subject matter covered. It is sold with the understanding that the publisher is not engaged in ren-
dering legal, accounting, or other professional advice. If legal advice or other expert assistance is
required, the services of a competent professional person should be sought.
— From a *Declaration of Principles* jointly adopted by a Committee of the American Bar Association
and a Committee of Publishers and Associations

MUSEUM OF SCIENCE STAFF:
Belinda Recio, Manager of Book and Product Development
Sally Ellison, Research Coordinator

SCIENCE REVIEWERS:
Lynn R. Baum
Ron Dantowitz
Olivier Schueller

*This book is available at quantity discounts for bulk purchases.
For information, call 1-800-872-5627 (in Massachusetts 617-767-8100).*

Visit our home page at http://www.adamsmedia.com

To my husband, Bruce Hrnjez,
for his love, advice, and high standards;

to my mother, Helen Schlesinger,
for her patience and support;

and to my friend Valerie Seabre,
for her unwavering enthusiasm.

Table of Contents

Acknowledgments

The Museum of Science would like to thank the following scientists for their contributions to this book:

Pamela Banning, MT, CLS
Lucía Muñoz Franco
Prof. Rainer Glaser
Dr. Laxman S. Kothari
Dr. M. Coleman Miller
Luisa Rebull
David S. Smerken, Ph.D.
Bette F. Weiss, Ph.D.

Introduction

When we tested the title of this book on potential readers, we found that some people thought it was a typographical error; others simply reversed the significant words of the title in their minds and read the title as *The Book of Questions & Answers*. Both reactions speak loudly to why we want to offer the world an answer-and-question book about things scientific.

Most people's experience with science is in school, where they learn that science is a collection of facts—things to be memorized. Science facts represent answers that scientists have arrived at after studying the world around them. Many of these answers are very interesting, and some are amazing. But to understand science as only a way to answer questions is truly to miss the point and to miss an opportunity to understand why scientists think that their profession is the most exciting and rewarding pursuit they can imagine. The intrigue and wonder of science are not so much in the answers it provides, but rather in the way it comes to those answers. In other words, the essence of science is asking questions.

Put another way, science is about curiosity. As children, we were insatiably curious about aspects of the world around us. We wanted to know how things work, what would happen if we changed something around, how various "parts" of the world interacted. Children ask much the same questions as scientists. Children attempt to find answers to their questions by observing closely, or by trying to alter what they're examining (e.g., What happens if I make a longer tail for the kite?), or, finally, if all else fails, by asking someone older and, supposedly, more knowledgeable.

Scientists do many of the same things children do when exploring the world. And some scientists will joyfully tell you that a profession that rewards them for following their curiosity is the best profession anyone could hope for. If you look a little deeper, however, you'll find some differences between the way scientists and children pursue and answer the questions their curiosity turns up.

Broadly, the difference is that scientists ask their questions in a more systematic way and are more discerning about the questions they ask. Not all questions are equally likely to yield useful information. Scientists control their variables very carefully, and they keep meticulous records of their observations and activities. So the scientist's curiosity is much more highly structured than the child's. But it is still, at heart, a curiosity that leads to questions that often produce answers.

So, if the process of science is one of asking questions that produce answers, why are we so adamant about calling this an answer-and-question book rather than a question-and-answer book? The answer lies in one further characteristic of the scientist's methodology. When a scientist posits questions, conducts experiments, or comes up with answers, these answers generate new questions. Every exploration, every experiment, unearths new questions, reveals mysteries, rekindles curiosity. This cycle of inquiry is the primary characteristic of science, and understanding it is central to our appreciation of why questions are so useful in science. The answers scientists find are never the last word. There is never a final list of "facts" or answers.

The process of science is not widely understood or appreciated. In fact, the public's misunderstanding of the nature of science has led to some confusion about the kinds of answers that science can and does produce. People frequently assume that science delivers "truths" or certainties about the world. But all that science can offer us is conditional or partial understanding. Science does not deal with certainties, only with probabilities. All scientific understanding represents the best current thinking on a subject. Any understanding is always subject to revision or replacement if other questions lead to new answers.

Therefore, when you read in the newspaper, as we did recently, that scientists have "discovered" billions of new galaxies, the proper response may not be "So that's what the early universe looked like," but "Ah, I wonder what else is out there, just waiting for a more powerful telescope to reveal."

With that spirit of curiosity and open-ended questioning, we offer you *The Book of Answers & Questions*. All of the essays—or "answers"— represent current scientific thinking on the subject. The essays are entertaining and informative, but the most intriguing part of the book may be the questions that follow each section. These represent the ongoing process of science in that every "answer" discovered by science only leads to a lot more questions. Needless to say, the questions that follow the essays are only a very few of all the questions raised by the essays. We tried to link—by referring you to titles and page numbers—these end-of-essay questions to other essays that offer insights, clues, and sometimes even answers to the questions. The linking questions add an element of nonlinearity that we hope will encourage you to navigate through the book based on your curiosity, rather than the sequential order of the essays. If, after reading the essay on octopus intelligence, you are intrigued by the question that follows—"How does memory work?"—you can turn to the page number listed and read the essay on memory. After reading about memory, you may follow a winding path through essays on the senses, chemistry of the brain, electricity, nuclear energy, and black holes.

Consider the end-of-essay questions a starting point for your own journey through the answers and questions of science. Keep a record of questions that come to *your* mind after reading each essay, and try to find your own answers. Our greatest hope is that this book will stimulate your curiosity. The essays should give you some sense of the enormous range of scientists' curiosity. They should also give you a better understanding of the process of science and the excitement available to us all if we learn to ask questions.

— JOHN SHANE
Vice President, Programs, Museum of Science

THE BIGGEST BANG ✓

In the big bang, our entire universe suddenly burst into being when a single point, smaller and hotter than we can imagine, exploded in a tremendous fury of inconceivable power and consequence.

The idea of the big bang is intimately connected with that of the expanding universe. In fact, it was the idea of the expanding universe that led scientists backwards, so to speak, to the big bang. In the 1920s, Edwin Hubble discovered that there are millions of galaxies in the universe, moving away from us at tremendous speeds. Further observations showed that the most distant galaxies were moving away from us the fastest, and nearby galaxies much more slowly. This is exactly what you would expect to see if the universe had begun in a supremely gigantic explosion—a "big bang." The fragments propelled fastest by the explosion would have had time to move farther away in space than the slower fragments. Hubble also discovered that the ratio between a galaxy's distance and velocity is constant (this value is known as the Hubble constant). This meant that at some time in the past—at the beginning of everything—all the galaxies in the universe were crammed into the same place at the same time. But how long ago was this celestial traffic jam, and the explosion that followed?

By measuring both the velocity and distance of various galaxies, it was a logical step for scientists to determine the age of the universe. Basically, they calculated how long the galaxies must have taken to reach their present location. The proposed answers vary somewhat— but not significantly, considering the huge size of the competing answers. Many scientists agree that the universe is from eight to twelve billion years old. Some researchers have estimated the age of the oldest stars in the Milky Way as fourteen billion years old. This causes some doubters to point out the paradox that the oldest stars may be older than the universe itself. But there is no real paradox:

scientists are constantly refining their data and theories, and eventually the numerical wrinkles may be ironed out. Part of the importance of determining the age of the universe is that scientists use that knowledge to help them understand how stars and galaxies formed.

What happened immediately after the big bang? First quarks and leptons, the constituent units of elementary particles, formed. In addition, the original single unified force separated into the four forces we know today—gravity, electromagnetism, and the strong and weak nuclear forces. And this was only in the first ten-billionth of a second! Next to form were the particles themselves, including protons, neutrons, and electrons. Then the first nuclei formed from protons and neutrons; and then the nuclei and loose electrons mixed together in a gas called plasma (not to be confused with the stuff in your blood). Finally, electrons, neutrons, and protons joined together in atoms, the familiar building blocks of the world as we know it today. Within an instant, this "stuff" had spread out to cosmic proportions.

Is there any evidence for the big bang? The first major evidence, discovered in 1964, was the existence of microwave radiation from deep space (yes, the same radiation that heats your coffee and turns your muffin soggy). If the universe came into being from a very hot point and has been expanding and cooling ever since, it should now be at a temperature of approximately –270 degrees Celsius, the same temperature at which celestial bodies radiate microwaves.

But what, you are probably asking, came before the big bang? Most likely, nothing, an unstable nothingness something like a vacuum. By chance, as is theoretically possible, a single dense particle of matter suddenly popped into being. And what is the end of the story? Scientists are divided on that one. The universe may continue to expand outward indefinitely, or it may one day collapse back onto itself. We'll just have to wait and see.

Because the universe is so large and so old, isn't it possible that life exists—or once existed—elsewhere? (See page 13: Extraterrestrial Intelligence)

If the entire universe collapses back into itself, will time run in reverse? (See page 3: Time's Arrow)

TIME'S ARROW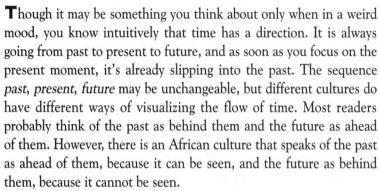

Though it may be something you think about only when in a weird mood, you know intuitively that time has a direction. It is always going from past to present to future, and as soon as you focus on the present moment, it's already slipping into the past. The sequence *past, present, future* may be unchangeable, but different cultures do have different ways of visualizing the flow of time. Most readers probably think of the past as behind them and the future as ahead of them. However, there is an African culture that speaks of the past as ahead of them, because it can be seen, and the future as behind them, because it cannot be seen.

However time is described, the essence of time's arrow is its irreversibility. Except for a few rare occurrences at the subatomic level, events cannot be reversed in time. This is made clear—and laughable—when a film is run backwards. Nothing looks plausible. There is no need to cry over spilled milk, because it climbs right back into the glass. Cars uncrash and fenders unbend. Snow rises into the sky, leaving the ground bare.

The irreversibility of time is related to entropy, the tendency for order to head toward disorder. The spilled milk and the dented car are less orderly than the milk in the glass and the smooth car. Even the snow falling through the sky is less ordered than the cloud from which it originated. Just as there is no physical law that says that disorder can't go toward order, there is no physical law that says that events can't go in reverse. There is just an extremely low probability—so incredibly small that it is unlikely to occur.

The probability that any collection of molecules that has gone from order to disorder will spontaneously move back toward order is so low because there are so many more disorderly arrangements than orderly ones. If you spill sugar from a sugar bowl, there are an infinite number of disorderly arrangements the sugar molecules can take on the table top. But the number of arrangements that could be considered orderly, such as a sphere or the letters of your first name, are so few that if you did see the sugar form a ball or spell out your name, you would probably think you were hallucinating.

Okay, so you're not likely to see time reversal at the dinner table. But how about on a larger scale, the largest possible—the universe? Although virtually all scientists believe that the universe was formed in the big bang and has been expanding ever since, some think it is likely that it will continue to expand forever, whereas others think that at some point it will collapse back in upon itself. What a film to run backwards! Such a collapse would be a form of time reversal. But there is another aspect to time's arrow—the human brain and human psychology. We have a subjective sense of time; most of us feel that it is moving forward. So even if we were to hang around long enough and survive the collapse of the universe, we would still experience the sequence of events as occurring in our familiar, forward-moving time.

If the universe might collapse back in on itself, could individual stars or galaxies do so as well? (See page 7: From Starlet to Has-Been: The Life Cycle of a Star)

Are some seemingly disordered states actually ordered in a way that is less obvious to us? (See page 81: Strange Attractors)

What is the difference between disorder and chaos? (See page 79: Chaos)

QUASARS

It's hard enough to imagine tiny Pluto, three and one-half billion miles from Earth. Quasars are billions of *light-years* from Earth (a light-year is the distance light travels in one year). Astronomers consider a quasar two billion light-years away to be nearby. Quasars, the brightest and most distant objects in the universe, can emit one hundred times as much energy as our entire galaxy, from a volume no larger than that of our solar system. Although astronomers are fascinated by these objects, both in their own right and for the information they shed on the vast amounts of material their light travels through to reach us, many of their ideas about quasars remain speculative. It is very difficult to study objects so distant from us in both distance and time, for we see the quasar not as it is now, but as it was billions of light-years ago, when the light that we see now left on its journey through space.

The brightest quasars are probably fueled by massive black holes, the final stage of collapse of the most massive stars. A black hole is so compressed that nothing remains of it but its incredibly powerful gravitational force, which captures anything that comes in the vicinity. As orbiting gas and dust approach, collisions between bits of gas and dust release energy. This rapidly spinning disc of light is what we see.

When astronomers discovered quasars in the early 1960s, they came up with the word *quasar* as a shortened form of "quasi-stellar radio source," because the quasars were strong radio sources. When they realized that most quasars are actually weak radio sources, they decided that *quasar* was really short for "quasi-stellar object."

In photographs, quasars look like stars. What tells astronomers that a quasar is not a star is its immense redshift. Redshift refers to a shift toward the red end (i.e., longer wavelengths) of the spectrum in the light of galaxies that are speeding away from us. Astronomers take the size of the redshift to be an indicator of distance. A minority of astronomers suggest that the immense redshift of quasars could be an effect of gravity or a Doppler effect of the movement of other galaxies (a Doppler effect is the apparent change in the frequency of a wavelength caused by the relative motion of two objects, as when the whistle of a passing train seems to change pitch as the train approaches and departs). Most astronomers, however, believe that quasars are indeed the most distant objects in the universe. Because quasars with very large redshifts are much more common than those with smaller redshifts, astronomers believe that quasars were more numerous in the early stages of the universe. The most distant—and most numerous—quasars, which are near the edge of the observable universe, formed when the universe was a mere infant, less than one billion years old.

Many galaxies have a bulge in their center. In some the bulge appears to be a quiescent (or dead) black hole, in others a quasar. Astronomers think that many galaxies may have gone through a quasarlike stage early in their development. Quasars may die when a galaxy reaches a certain size, because as more gas is caught up in the stars, less is left to fuel the quasar. Most of the quasars that once burned in the universe are now dead.

When quasars were first discovered, the gap in luminosity between them and ordinary galaxies was so huge that some astronomers preferred to question accepted physical laws rather than accept that quasars were as bright as they appeared to be. Recently, however, astronomers have discovered galaxies whose brightness falls between that of ordinary galaxies and the incredible brightness of quasars. They call these unusual galaxies "active galaxies."

The light of a quasar is like a message from the earliest days of the universe that finally reaches us after these billions of years. Our interpretation of the message is constantly changing.

How can light "shift"? Is it slowing down? Is the light itself changing or just our experience of it? (See page 40: The Doppler Effect)

FROM STARLET TO HAS-BEEN: THE LIFE CYCLE OF A STAR

A star is born by chance—not, as in Hollywood legend, when a producer discovers one sitting on a drugstore stool, but when bits of matter in the diffuse clouds of outer space happen to glob together. Then gravity joins in on the work begun by chance. Like all objects, the glob of star stuff exerts a force of gravity. Gravity draws more material into the glob, which, of course, then exerts an even stronger force of gravity. Eventually gravity causes the now-massive glob to contract in upon itself. The story would end here with a black hole, were it not for the fact that as the glob of matter contracts, heat and pressure build in its center. As the temperature and pressure continue to rise, nuclei move faster and faster until finally they join together and nuclear fusion takes place. Now the pressure of the newly formed star is able to counteract the force of gravity. The nuclear energy finally escapes from the mass and travels through space in the form of electromagnetic radiation—this is what we see shining. The star rises from her stool to take on Hollywood.

Gravity, which gives a star its beginnings and holds it together, is also its downfall. Throughout its life cycle the star is fighting off the total collapse that gravity threatens. Its battle with gravity determines a star's changes from one stage of stellar evolution to another. Because these stages take place over such long periods of time, scientists cannot directly observe the change from one stage to another. Instead, they use statistical evidence to determine the time span of the different stages. In other words, the more numerous the stars at a particular stage, the longer scientists assume that stage to last.

Once a star begins nuclear fusion and stabilizes, it enters a long period during which it is called a main-sequence star. The more massive a star is, the more fuel it must burn to counteract the force of gravity; thus, the more brightly it burns, and the shorter is its life span. Our Sun, which is a medium-sized main-sequence star, has been burning brightly for approximately five billion years and has another five billion to go before it needs to check its fuel gauge.

As a main-sequence star begins to deplete the fuel in its center, gravity causes it to start to contract again, and contraction again causes temperatures to rise. Though the core fuel is depleted, nuclear reactions now take place in the shell surrounding the core. While the core contracts, the layers outside the shell expand. As the size of the star increases, the outer layers cool off and the star's color turns from yellow to red. The star is now called a red giant. Because there are fewer red giants than main-sequence stars, scientists assume that their life span is shorter.

At some point, the red giant runs out of energy and begins to contract again. With less massive stars, the electrons of the core reach a point where they refuse to be compressed any further. The star is newly stabilized: gravity pushes in and the electrons push out. Although they have no fuel, these stars, which are called white dwarfs, shine for a long time as they cool off. Our Sun will probably end up as a white dwarf. In more massive stars, the electrons in the core cannot resist the force of gravity. They are forced together with protons to form neutrons, and eventually the star stabilizes as a neutron star. Neutron stars are so dense that one with the mass of our Sun would have a radius of only about six miles. If the star is still more massive—so massive that the neutrons cannot resist the force of gravity—it totally collapses in upon itself to become a black hole.

With the most massive stars of all, in a matter of only hours the cool outer layers contract into the center and heat up so rapidly that a tremendous nuclear explosion ignites, bursting the star into pieces. This event, called a supernova, is quite rare, occurring only two or three times per century in each galaxy. After a few days of spectacular fireworks, the supernova ends up as either a neutron star or a black hole.

Most stars do not end up as neutron stars or black holes, but as white dwarfs. Like many Hollywood stars, white dwarfs simply hang around for a long time until they fade.

Why do only the biggest stars turn into black holes? (See page 15: Black Holes)

THE ASTEROID BELT ✓

Debris. The asteroid belt is basically a band of rocky debris that orbits the Sun in the gap between Mars and Jupiter. But that celestial debris can tell astronomers a great deal. Astronomers once speculated that the asteroids were the fragments of a shattered planet. However, they now know that the asteroids differ from each other too much in chemical composition to have originally belonged to the same object. Also, to blow a planet apart, an explosion would have to be powerful enough to overcome the binding force of the planet's own gravity. That's unlikely. Astronomers now think that the asteroid belt consists of fragments that never successfully came together to form a planet.

Asteroids are not very big. The largest, Ceres, is about 620 miles (1,000 kilometers) in diameter. Together, the thousands of asteroids in the belt would have only 1/200th the mass of Earth. Astronomers determine the size of asteroids by measuring the amount of sunlight they reflect and heat they radiate. Astronomers determine the chemical composition of asteroids through spectroscopy, which allows them to identify the characteristic spectra of different materials.

Although it is called a belt, the asteroid belt actually consists mostly of empty space. Even the largest asteroids rarely collide. When they do, they shatter into many smaller fragments that speed away from the original asteroids and take up their own orbits around the Sun. Astronomers call the "parent" asteroids and the newly formed fragments an asteroid family. Because members of a family often end up orbiting at different places in the asteroid belt, astronomers look at the specific shape and orientation of an asteroid's orbit to determine its family alliance. Of course, astronomers can mistakenly group together asteroids that are, in fact, unrelated. This possibility gives rise to disagreement as to the actual number of asteroid families. Some astronomers consider only about twenty families as giving strong enough evidence of their common origins. Others rate as many as one hundred groups as true families. The question is important not, as among humans, to decide who gets invited to a wedding or bar mitzvah, but to help astronomers understand the formation and nature of asteroids.

If, as many scientists think, an asteroid struck Earth, killing off the dinosaurs, could we someday suffer the same fate? It's possible, but highly unlikely. Astronomers estimate that Earth is struck about once every 100,000 years by an object about 0.6 miles (1 kilometer) in diameter and about once every 50 million years by an object about 6.21 miles (10 kilometers) in diameter. Though it probably requires a 10-kilometer object to kill off an entire species, whether of the dinosaur or human variety, a collision with an object even one kilometer in diameter would have more energy than all of the nuclear weapons now stocked on Earth.

Could the asteroids in the asteroid belt ever try again and successfully form a planet? No. First of all, all their mass together would not be enough to form a respectably sized planet. But more significantly, Jupiter's gravitational field constantly disrupts the motions of the asteroids, preventing them from coming together into a clump large enough to adhere together from its own gravity. But if they're disappointed with their status in the solar system, the asteroids might at least take consolation in the fact that some astronomers refer to them as planetoids.

Could Earth get slammed by the "big one" five minutes from now? (See page 75: Probability)

Did any dinosaurs survive the impact? (See page 115: Why There Couldn't Be a Lone Dinosaur or Two Alive Somewhere)

If an asteroid did kill the dinosaurs, then why did any life survive on Earth? (See page 107: Evolution: Gradual or Sudden?)

Could asteroids have weather? (See page 104: Weather Systems)

REFLECTING TELESCOPES

Most people think of a telescope as a tubular instrument that allows one to see faraway objects. Technically, however, a telescope is an instrument that collects radiation from distant objects. In addition to visible light, that radiation includes infrared light, ultraviolet light, X rays, gamma rays, cosmic rays, neutrinos, and radio waves. Most astronomical telescopes are optical telescopes—telescopes based on light waves from the visible part of the spectrum.

There are two basic categories of optical telescopes—refracting and reflecting. Basically, a refracting telescope uses lenses and a reflecting telescope uses mirrors. Galileo made the first refracting telescope for astronomical use in 1609. The lenses of a refracting telescope are ground to refract—that is, bend—the light waves reaching the telescope from a distant object. The light waves are collected by the first lens, called the converging lens, which causes them to bend. This image is then magnified by a second lens, called the eyepiece. A significant problem with refracting telescopes is called chromatic aberration. This refers to the fact that light waves of different lengths, e.g., red and blue, focus at different points. If the red is in focus, the blue is smeared, and vice versa. Another problem with refracting telescopes is that there is a limit to how large the lenses can be. Because they must be transparent, they can be supported only around their edges, where they are thinnest and most fragile. The largest refracting telescope in the world, which was completed in 1897, is 40 inches (1 meter) in diameter.

Galileo was a pretty smart guy. So was Newton. In 1668, he made the first reflecting telescope. It used a concave mirror to collect and focus the light; a second, smaller mirror placed at a 45-degree angle to the main one reflected the light onto the eyepiece, which magnified it. Today's reflecting telescopes are much larger. Because the light bounces off a surface layer of aluminum instead of going through the mirror, the mirror itself does not have to be transparent. Thus, it can be supported at any place on its surface, allowing it to be built much larger than the lens of a refracting telescope. The larger the telescope, the more light it collects—a significant factor when you're observing objects many light-years away. The largest (reflecting) telescope in the world today, the Keck Telescope in Mauna Kea, Hawaii, measures 400 inches (10 meters) in diameter. Telescope designers are not limited to using a single large mirror. Some telescopes, including the Keck, use numerous small mirrors that are kept in focus by computers. Mirrors have another advantage over lenses—they don't distinguish between wavelengths when reflecting light. Thus, there is no problem with chromatic aberration.

For most of us the word *see* means to register an image with our own eyes, perhaps with the aid of glasses or contact lenses. But many scientists spend their days peering through the lenses and mirrors of their telescopes to "see" distant galaxies.

Can we see any black holes with our advanced telescopes? (See page 15: Black Holes)

EXTRATERRESTRIAL INTELLIGENCE

When scientists speak of extraterrestrial intelligence, they usually mean life forms that look nothing like us physically and whose civilization is very unlike ours, except for one thing—they are technologically developed. In a sense, it's not our Picasso, Beethoven, or Confucius that makes us intelligent—it's our CD players and microwave ovens. Not that scientists wouldn't be interested in how life forms on other planets work and play (if that's what they do), but unless they have the technology to communicate with Earth, we'll never know anything about them.

The issue of whether or not to search for extraterrestrial intelligence has two parts—whether such civilizations exist, and whether we can communicate with them if they do. Scientists are of two—often vociferous—opinions as to the likelihood of the existence of extraterrestrial intelligence. The skeptics say that life on Earth evolved under a very precise set of chemical and climatic conditions that are unlikely to be duplicated elsewhere. In addition, humans were the result—so far—of a long process of evolutionary mutation and adaptation that could easily have come out differently.

Those in the other camp adhere to the "mediocrity principle," which says that there's not really anything so special about our solar system, planet, or species. In the Milky Way Galaxy alone, there may be about forty billion stars similar to our Sun. If, as most astronomers believe, the formation of planets is common when stars condense from clouds of gas and dust, then there may also be billions of other planets. If only a very tiny percentage of them have conditions similar to those on Earth, that would leave a lot of candidates for planets able to support intelligent life. According to biological evolutionists, the possibility that extraterrestrials would follow the same evolutionary path as humans is extremely unlikely. But because intelligence—which enables one to process information and manipulate the environment—gives such a survival advantage, the trait would probably develop, no matter how different the life forms might otherwise be.

The search for extraterrestrial intelligence (SETI) began in earnest with the development of large radio telescopes, because radio waves are the most likely way for other civilizations to communicate with us. One advantage of radio waves is that they travel at the speed of light, no small consideration when you want to send a message to someone tens of light-years away. By the late 1980s, scientists had conducted some forty-five radio searches (you can probably guess that the results were negative). Then, in October 1992, on Columbus Day, NASA embarked on a ten-year search mission involving telescopes stationed around the world. The $100 million project had two components. One, a so-called targeted search, was to use highly sensitive radio telescopes to scan those areas of the sky where scientists thought a payoff to be most likely. The other was to use less sensitive instruments to scan the entire sky . . . just in case. But a year after the project began, Congress cut off its funding. The nonprofit SETI Institute, which was working as a NASA contractor, sought private funding. Though it was able to raise money, the full-sky search, which was housed at NASA's Jet Propulsion Laboratory, stopped. According to SETI officials, within a few years it may not be possible to conduct a full-sky search from Earth because of interference from the signals of cellular phones, satellites, and the like.

SETI scientists are following an intellectual line of questioning that goes at least as far back as Copernicus. Then the question was, is Earth the center of the solar system? Now the question is, are we the only intelligent beings in the universe? Of course, on a more practical level, the question is, is anybody out there trying to get through to us? And if so, how long are they willing to wait for a response?

If scientists discovered radio evidence of extraterrestrial civilization, would that evidence come from the past or present? (See page 5; Quasars)

BLACK HOLES

Many objects in the universe are too large, hot, or distant to even begin to visualize. But some objects, such as black holes, are almost too strange to imagine. A black hole, which is the final stage of the gravitational collapse of a very massive star, is so compressed that it consists of nothing but its gravitational force. And what a force! When something is sucked into a black hole, it's on a one-way ticket; there's no return trip. But it doesn't much matter, because anything so unlucky as to go down a black hole would be crushed to what astronomers call infinite density. You can't get crushed any smaller than that.

When astronomers say that nothing comes out of a black hole, they mean nothing—not even light. This poses a problem when it comes to observing a black hole, for the primary way astronomers identify and measure celestial bodies is by the light they emit or reflect. Because astronomers can't identify black holes directly, they speak of "candidates" for black holes. They detect the presence of a black hole by observing its gravitational influence, such as the bending of light waves in close vicinity. Black holes that result from the collapse of one of a pair of stars, called binary stars, provide additional clues. As the black hole draws material from its companion star, the material swirls around in what astronomers call an accretion disc. As the material swirls, it heats up, emitting X rays. The constellation Cygnus, in the Milky Way, contains a star system called Cygnus X-1, which consists of an X-ray–emitting supergiant star that is revolving around an invisible companion. Cygnus X-1 is one of the strongest candidates for a black hole.

A black hole is so dense that one with the mass of five of our suns would have a radius of only about 12 miles (20 kilometers). The entire collapsed mass is surrounded by a boundary called an event horizon. It marks the point of no return for anything unlucky enough to go past it.

If you have a taste for paradox, then black holes are for you, for some astronomers think that these darkest of objects may be the source of quasars, the brightest and most distant objects known. In a single second, a quasar can emit as much energy as our Sun produces in ten thousand years, even though the quasar is only a few times the size of our solar system. Astronomers speculate that as orbiting gas and dust approach a black hole, bits of gas and dust collide with other bits, releasing energy. With black holes a hundred million times more massive than our Sun, this released energy may be the light of quasars.

Stars are born and die, and the most massive end up as black holes. But then once a black hole, always a black hole. As a black hole draws more matter into itself, it simply grows larger. None of those one-way tickets can be exchanged for round-trip.

If the most massive stars eventually end up as black holes, then what happens to the not-so-massive stars?
(See page 7: From Starlet to Has-Been: The Life Cycle of a Star)

Could black holes be full of dark matter? (See page 19: Dark Matter)

SUPERSTRING THEORY

Recently, some scientists have taken to describing the universe in terms of strings. Scientists use visual images such as strings to help them think about things otherwise too abstract to grasp. But that doesn't mean that the universe is composed of tiny waving strands of spaghetti, or dental floss. To begin to understand superstring theory, it is useful to look at two things: the size at which these hypothetical strings exist and the motivation for scientists to come up with such peculiar entities.

First, size: we're talking small, very small. Hypothetically, a single string is to an atom as an atom is to Earth. Second—and a bit more complicated—is the motivation for this theory. The most ambitious, most exalted quest in science is scientists' search for the Holy Grail of physical theory—a so-called Grand Unified Theory, or Theory of Everything. Physicists' theories describe the world on two levels: the macroscopic and the microscopic. Einstein's theory of general relativity, which describes the universe as a continuous space-time fabric in which the presence of mass creates the gravitational field, applies to the universe on the macro, or everyday, level. On the molecular and atomic level, quantum theory describes the behavior of matter and energy in the form of very, very tiny discrete, or separate, entities whose locations scientists describe in probabilistic terms. The forces physicists are trying to unite operate on these two levels.

There are four fundamental forces. Gravity is called the weak force. The other three forces, which are described by quantum theory, are electromagnetism, the strong nuclear force, and the weak nuclear force. Until the development of string theory, and some more recent offshoots, physicists had no way to combine the theory of gravity and quantum theory. String theorists suggested a bridge between these two realms called a graviton, a theoretical particle that transmits gravity in much the same way that the particle called a photon transmits light.

The first string theory, developed in the late 1960s, had the daunting drawback that it worked only in a universe of twenty-six dimensions, as opposed to the four of the theory of general relativity (length, width, height, and time). Then in the 1970s, scientists combined string theory with what is called supersymmetry to come up with superstring theory. Supersymmetry is a form of symmetry (or interchangeability) between particles that make up matter and particles that transmit forces. In superstring theory, the different vibrations of tiny stringlike particles give rise both to the particles that make up matter and to all of the four forces. Science writers often compare these vibrations to those of violin strings that produce the many different notes. Superstring theory provided some relief to scientists by reducing the required number of dimensions from twenty-six to ten. What are the extra six? Hard to say. They may be dimensions that were present at the instant of the big bang but then immediately curled up into some configuration currently invisible to us.

Superstring theory has garnered a lot of serious investigation, as well as a lot of popular press. It's not the only fibrous theory in the works, however. In the 1980s, another scientist came up with loop theory. Unlike superstring theory, which attempts to show how gravity and the other three forces are all manifestations of a single force, loop theory merely (!) attempts to explain how gravity acts on the particles of the quantum world. According to loop theory, rather than a smooth continuous fabric, space-time may be composed of tiny linked loops. Thus, loop theorists were the first to be able to describe space-time in terms of discrete units.

Strings, loops . . . what next? The four-dimensional world is hard enough for us to understand—never mind the strange multidimensional metaphors constructed by scientists. The important thing to remember is that the strings and loops in these theories are just visual tools to help scientists imagine and explore the most abstract of realms. Paper clips, anyone?

If matter is made of particles, what are forces made of?
(See page 37: The Electromagnetic Spectrum)

DARK MATTER

All the talk of galaxies, red giants, white dwarfs, supernovas, black holes, and so on gives the impression that astronomers pretty much know what the universe is all about. But to astronomers' chagrin, all these celestial objects make up only 10 percent of the universe—90 percent of the matter in the universe is unaccounted for. One name for the other 90 percent is the *missing matter*. Of course, the missing matter of the universe isn't really missing; we just don't know where or what it is.

Why do astronomers think that there's another 90 percent to the universe? Basically, to explain the motion of the galaxies. Galaxies rotate at very high speeds, and gravity is the only force that could keep the galaxies from flying apart as they rotate. Gravity is proportional to mass, which is the amount of stuff in an object. To account for the amount of gravity necessary to keep the galaxies from flying apart, astronomers must assume a certain amount of mass in the universe. When they add up all the known objects in the universe, it adds up to only 10 percent of the required mass. Hence, the assumption of the missing mass.

Another name for the missing matter is dark matter. Astronomers call the missing mass dark matter because it doesn't seem to absorb or give off any type of radiation. Of course, if it did, then it wouldn't be the missing matter. To further complicate matters, astronomers think that 20 percent of the dark matter is hot dark matter, and the other 80 percent is cold dark matter. Some researchers think that the hot dark matter may consist of neutrinos, which belong to the group of elementary particles called leptons. Although scientists used to think that neutrinos had no mass, in 1994 physicists at Los Alamos announced experiments showing that neutrinos may have enough mass to account for 20 percent of the dark matter. Astronomers call the cold dark matter "cold" because they think that it probably emerged from the big bang at speeds much less than that of light. In 1995, some physicists came up with evidence for another type of subatomic particle that could account for at least some of the cold dark matter. They call it WIMP, for weakly interacting massive particle. In 1996, a British team of astronomers reported evidence that the burned-out stars called white dwarfs may account for 50 percent of the dark matter.

Other candidates for the dark matter have been put forth and dismissed. For a while, some astronomers thought that black holes might be the source of the gravity produced by the dark matter. Other astronomers thought that the dark matter might simply be ordinary stars that weren't bright enough for telescopes to detect. But the Hubble Telescope found fewer faint stars than astronomers expected. Findings from the Hubble Telescope also indicate that there seems to be a lower limit of mass to stars; at least in the Milky Way, there don't seem to be any stars with less than 20 percent of the mass of our Sun.

The dark matter may not be missing, but it is definitely mysterious. The truth about what it is may turn out to be completely different from current lines of thinking. But for those who believe that science is more about questions than answers, the mystery is part of the fun.

What are the elementary particles? (See page 23; Quarks)

ENTROPY

Entropy can be summarized by the fate of poor Humpty Dumpty: "All the king's horses and all the king's men, couldn't put Humpty Dumpty together again."

Entropy is a measure and a direction. It is a measure of disorder, and it's the direction toward which all things are headed: more disorder. If a jar contains red, blue, green, and yellow marbles mixed up randomly, the level of color entropy is high, so high that it is called maximum entropy with respect to marble color. If you arrange the marbles in layers of red, blue, green, and yellow and then shake the jar, in a very short time the marbles will reach a state of maximum entropy. That is, as you shake the jar, the collection of marbles is headed in the direction of disorder. Disorder is also described in terms of probability; the most disordered state is the most probable. When you shake the jar, the most probable, or likely, state is the one in which the marbles are mixed randomly. It is very unlikely that the marbles will return to their original layered state, so unlikely that if you shook the jar for a billion years you would probably not see it happen.

Disorder is much more easily and quickly achieved than is order. Think how long it takes to build a house and how quickly it can be demolished; how long it takes to sew a seam and how quickly it can be ripped apart; how long it takes to arrange a salad prettily on a plate and how quickly it can be scraped into the garbage.

If we were to say that the universe has desires, we would say that it wants to relax into disorder. So how, you may ask, is order ever achieved? The answer is, at the expense of disorder somewhere else. When scientists study problems, they often think in terms of closed systems. A closed system, such as the neat row of books on your shelf, may have a definite order. But how did it get that way? You, or some other external agent such as a sibling or spouse with neater habits than yours, had to stand them in that row. To do that, they expended energy, creating disorder: orderly foodstuffs were broken down in the body, excreted, and burned off in the act of work.

The most highly ordered systems are those of life. A living creature has to be exquisitely ordered to survive. In fact, the way scientists recognize cancerous cells is by the disorder they see on a slide. The more disorder in the cells, the more certain is the diagnosis of cancer.

The nature and implications of entropy are stated in the three parts of the famous second law of thermodynamics. The first part says that heat energy always flows from a hot object or place to a cold one. If you leave your door open in the winter, the heat will escape. If you leave the door to your air-conditioned house open in the summer, the cold does not escape—the heat comes in.

The second part says that no machine, such as an engine, can convert energy to work with total efficiency. Some energy is always lost by dissipation into an area of lower temperature. In a car, for example, heat is radiated from the engine and expelled through the exhaust pipe.

The third part says that all systems tend toward disorder with the passage of time. Scientists refer to this as the arrow of time. Think of watching a film of a bucket of milk tipping over and spilling. If you run the film backwards to show the milk returning to the bucket and the bucket returning to an upright position, you may be highly amused, but you also will know by the improbability of the sequence that the film is being run backwards.

The largest closed system, in case you haven't already guessed, is the universe itself. And, yes, it is heading toward greater entropy. The big bang created a temporary—though very long, by our standards—drop in entropy when the universe exploded into being. As the universe continues to expand, it drifts toward greater disorder. If it expands forever, it may reach a burned-out state that, like Humpty Dumpty, can't be put back together again.

Is the backward flow of time impossible, or merely improbable? (See page 3: Time's Arrow)

QUARKS

Ever since humans began to think about the world around them, they've wondered about the units that make up matter. As early as the fifth century B.C.E., Greek philosophers described small, homogenous parts they called atoms. Most of us know that matter is made up of atoms, which can be broken down into protons, neutrons, and electrons. Scientists, in their immodesty, went so far as to call the parts of the atom the elementary particles. But the story doesn't stop there. Now scientists believe that the so-called elementary particles can be further divided into two kinds of particles, called leptons and hadrons.

Leptons, which include electrons and a few more exotic particles, appear to have no internal structure. Hadrons, which include protons and neutrons, have a complex internal structure. Hadrons are built of quarks; thus, the two elementary particles are really leptons and quarks. Unlike electrons and protons, which have opposite and equal electric charges, quarks have fractional charges, either a two-thirds positive charge or a one-third negative charge. Quarks have antiquarks, which carry the opposite fractional charge. An antiquark is a type of antiparticle. All antiparticles have the same mass as their opposite particle and an equal but opposite amount of some property, such as electrical charge.

Since the complex hadrons consist of quarks, quarks are central to scientists' understanding of the subatomic world. The scientist who devised quark theory, Murray Gell-Mann, originally called them *quorks*. But one day while paging through James Joyce's *Finnegans Wake*, he came across a passage with the word *quark* in it and liked it for its funny sound and multiple meanings (if you're curious about Joyce's quarks, read *Finnegans Wake*). Physicists seem to like whimsical terminology, for they went on to describe the varieties of quarks by what they call flavors. There are six flavors of quark—up, down, charmed, strange, top, and bottom. According to theory, all six flavors were created shortly after the big bang. Only the up and down quarks, which make up all protons and neutrons, survived in nature. Although the other flavors of quarks disappeared from the observable universe, scientists have been able to create them through very high-energy collisions in particle accelerators. Though all of the other quarks had been identified by 1977, the top quark (sometimes called *t* for truth) remained elusive until March 1995, when its identification was announced to the world with great fanfare.

Just in case six flavors are not enough variety for you, each flavor comes in three colors—red, green, and blue. The antiquarks come in the complementary colors—cyan, magenta, and yellow. Thus, there are a possible eighteen quarks and eighteen antiquarks. The rule of quark artistry is that any combination must give white, either by combining the three primary colors, or by combining a color with its complement. Keep in mind that the terms *flavor* and *color* have nothing to do with what we mean when we talk about ice cream or living room walls. The terms are simply tools to help scientists think about complex phenomena.

Numerous questions about quarks remain. Though quark theory has been firmly established, no one has seen an isolated quark—in the sense that scientists "see" subatomic particles, usually by some kind of trail left by the particle. A lone quark should be easy to detect because of its fractional electric charge. It is also possible that there are more than six flavors of quark. And, most interesting, it is possible that scientists have still not reached the true elementary particles; the quark itself may consist of even smaller units.

How were the quarks created in the big bang? (See page 1: The Biggest Bang)

What happens when a quark collides with an antiquark? (See page 27: Antimatter)

NEUTRONS ✓

Of the three subatomic particles—the electron, proton, and neutron—the neutron is the most versatile and is of vast scientific and technological importance. It is the only one that is without an electric charge, that is, neutral; hence the name *neutron*. Like a person with an unassuming personality who easily handles social situations ranging from cyberspace dating to science club meetings, the neutron uses its neutrality to venture where the electron and proton cannot go. Its uses range from the mundane to the terrifying, and it is the stuff of dying stars.

Neutrons and protons are bound together in the nuclei of atoms and surrounded by clouds of electrons. Scientists produce isolated neutrons by forcing apart nuclei. After approximately fifteen minutes, an isolated neutron decays into an electron, an antineutrino, and a proton. (Although one characteristic of an antiparticle is that it has the opposite charge of its particle, since neutrons and neutrinos are neutral, so are antineutrons and antineutrinos.)

Scientists like to use neutrons to examine various materials at the molecular and atomic level. The technique they use is called neutron scattering. First, the scientists produce beams of neutrons through fission, the forcing apart of nuclei. They then shoot these beams at the material they want to study. X rays, which are more familiar to most people, interact with the electron clouds of the atoms they are "sent" to examine. Neutrons, which basically ignore the electrons, can penetrate farther into the nuclei of a material. Because neutrons are much smaller than electron clouds, many neutrons in a beam sail right through the sample without hitting anything. But those that hit nuclei bounce off and are scattered in different directions. The geometry of the nuclei the neutrons hit determines the angle at which the neutrons scatter, and this information is "read" by the scientists. If the collisions are inelastic—that is, if the neutrons don't bounce off the nuclei—scientists get equally valuable information from the energy transfer that occurs when the neutrons hit the nuclei.

Neutrons have the leading role in nuclear fission—both the controlled kind used in nuclear reactors and the uncontrolled kind used in the atomic bomb. In nuclear fission, a neutron hits the nucleus of a material such as uranium. This causes the nucleus to split apart, emitting two or three neutrons. The amount of energy emitted is equivalent to the difference in mass between the original nucleus and the resulting fission products. The neutrons that are spewed out go on to hit other nuclei, causing a chain reaction. Once the chain reaction reaches a particular "critical mass," it goes out of control and there is an explosion. Atomic bombs are made with two separate masses of material that together exceed the critical mass. The two masses are surrounded by conventional explosives. When these explosives are set off, the two masses are pushed together, causing the atomic explosion.

The importance of neutrons is not limited to nuclear reactors and atomic bombs. As a star reaches the end of its life cycle, it becomes so compressed that the negatively charged electrons are pushed into the nuclei, where they team up with the positively charged protons to form neutrons. Thus, eventually the star consists almost solely of neutrons. It is now a neutron star, with a density far greater than anything else in the universe except black holes . . . not bad for tiny particles without even an electric charge.

What about the beginning of a star's life cycle—how is a star formed in the first place? (See page 7: From Starlet to Has-Been: The Life Cycle of a Star)

Antimatter

The word *antimatter* conjures up science fiction images of invisible worlds lurking behind our everyday reality. To the physicist, antimatter is simply matter composed of antiparticles. For each subatomic particle, there is an antiparticle—a particle with the same mass and an equal but opposite amount of one or more properties, such as charge. An electron, for example, has a negative charge; the electron's antiparticle, the positron, has a positive charge. In 1928, the French physicist Paul Dirac predicted the existence of positrons, and in 1932, an American physicist named Carl Anderson first detected them in a spray of cosmic rays (cosmic rays are high-energy particles that reach Earth from space).

For scientists, the problem with antiparticles is that they don't hang around for very long. As soon as an antiparticle comes in contact with a particle, the two are immediately annihilated, producing pure energy in the form of radiation. This poses no great risk when it comes to individual particles and antiparticles meeting their doom, but if a large glob of antimatter were to meet an equally large glob of matter—which it could hardly avoid doing—it would set off an extremely powerful explosion. Some scientific visionaries have even suggested the possibility of one day using antimatter as a fuel for interstellar travel.

Physicists have been able to produce antiparticles since the 1930s, using high-energy accelerators. But they haven't yet been able to produce a single atom of antimatter. The obvious candidate to try is antihydrogen, since hydrogen has only a single electron orbiting a single proton. Now, however, physicists are getting close to mating an antiproton with a positron. The trick is to get them close enough to form a stable antiatom.

As exciting as it will be to form atoms of antihydrogen, physicists will only be able to verify their success when the same atoms are destroyed. The evidence that they have been destroyed will be the evidence that they existed. In the meantime, scientists are working on the development of containers with which to trap antimatter and protect it from annihilation. The containers, which would resemble large Thermoses, would have a very high vacuum inside, and magnetic and electric fields to keep the antimatter away from the container walls.

One reason physicists are excited about creating antimatter is that they will be able to find out whether it differs from matter in its behavior. Two questions they are particularly eager to answer are whether the spectrum of an antiatom is the same as that of its atom, and whether gravity acts differently on antimatter from the way it acts on matter. Antimatter may also have potential practical uses in manufacturing and medicine. In medical imaging, for example, anti-matter would require smaller doses of radiation than current X rays and CAT scans.

As scientists go to all this trouble to produce atoms of antihy-drogen, science fiction writers speculate about entire antigalaxies and antiuniverses. The only real argument in favor of an antiuni-verse is that as long as it didn't come in contact with any matter, it would be stable. Most scientists, however, think the existence of antigalaxies and antiuniverses unlikely. Right after the big bang, the resultant fireball probably contained all the known particles, with equal amounts of matter and antimatter. There is a slight difference, however, in the decay rates of matter and antimatter. Thus, as matter and antimatter annihilated each other, there would have been a slight excess in the amount of matter. As the universe con-tinued to expand, that slight difference grew and grew. Perhaps, somewhere an antiscience fiction writer may be speculating about a universe just like ours.

If particles of antimatter have opposite charges, do antineutrons (neutral antiparticles) exist? (See page 25: Neutrons)

INERTIA ✓

*I*nertia is one of those scientific words that make it into our everyday language while retaining only a shred of their original meaning. Most people think of inertia as what makes it hard to get out of bed in the morning or to rise from the couch while watching TV in the evening. But it's also what makes it hard to stop when you're ice skating. Inertia made its first conceptual appearance in the second of Sir Isaac Newton's famous three laws of motion. Inertia is defined as a body's resistance to change in motion. In other words, with no external force acting upon it, a body at rest remains at rest, and a body in motion continues in motion in a uniform straight line. (Though Newton was English, he wrote in Latin, so you may come across different translations of his work.) Part of the power of Newton's work was its universality; the law of inertia applies to everything from a marble rolled across a floor to the planets in their orbits (gravity keeps them circling around the Sun instead of flying off in a straight line).

In case things seem too simple, we'll complicate them a bit by introducing the term *mass*. Mass refers to how much stuff there is in an object, e.g., how many atoms in a candy bar. Many people confuse mass with weight. But weight, which measures the force of gravity on an object, changes with changes in gravity—thus the weightlessness of objects in outer space. The mass of an object is the same no matter where the object is, for the amount of stuff in it doesn't change with location. Mass is also the measure of a body's inertia. In other words, the mass of a body refers to how much force it takes to move the body if it is at rest, or to change its motion if it is in motion. A refrigerator has more mass than a candy bar; therefore, it takes more energy to move it.

Acceleration is another word whose scientific meaning is much broader than its everyday one. Most people think to accelerate means to speed up. Actually, acceleration refers to any change in motion—speeding up, slowing down, or changing direction. According to Newton's first law, all of these forms of acceleration require the application of force. If you measure the amount of force required to slow down a speeding train, you can use that measurement to calculate the train's mass.

A tragic example of inertia is too often provided by a person driving a car with the seat belt unfastened. For example, if there is an accident and the car hits a brick wall, the force applied by the wall causes the car to come to a stop. But inertia causes the unfortunate driver to continue moving at the car's rate of motion until acted upon by a force—usually applied by the steering wheel or windshield.

Will inertia cause the universe to continue expanding? (See page 1: The Biggest Bang)

THE PHOTOVOLTAIC CELL

Although we may not realize it, when we pull open our curtains to let in light and warmth, we're using solar energy. When we design our house to make the most efficient use of light and warmth from the Sun, then we're using solar energy more deliberately. When we convert sunlight to electricity to run our watches, calculators, or traffic lights, we're using sunlight as a fuel. Using sunlight as a fuel, however, is almost like not having to use fuel at all—it's abundant, free for the taking, and endlessly renewable for as long as the Sun shines (a few billion years more, at least). The most common way to use sunlight as a power source is in the photovoltaic, or solar, cell. The word *photovoltaic* comes from *photo*, for "light," and *volt*, the measure of electric potential.

The operation of a photovoltaic cell takes place in three stages: the absorption of sunlight; the separation and movement of positive and negative charges; and the transmission of the charges, in the form of an electric current, to outside the cell.

The amount of sunlight absorbed depends, in part, on the available sunlight. An electrician's license is not necessary to evaluate the sunlight; a simple walk outside will usually do. The cell's ability to absorb sunlight is also affected by the materials used to make the semiconductor. (A semiconductor is just that—halfway between a conductor and an insulator of electricity.) When the light is absorbed by the semiconductor, it separates into negatively charged electrons and positively charged holes. The structure that makes the photovoltaic cell possible is the electrical junction, which can be described as a boundary between two dissimilar substances. In the photovoltaic cell, the two dissimilar substances are often a metal and the semiconductor. When light is absorbed by the semiconductor, it produces electrical charges in the cell. The electrical junction allows these charges (electrons) to cross over from one of the dissimilar materials to the other, but not to return. This separation of charges creates what is called a potential, or voltage difference between the materials. The voltage allows a current of these electrons, or electricity, to flow through and leave the cell through an external circuit.

Earth is not the only place where photovoltaic cells are used. No, we do not have evidence of their use by Martians; however, they do generate the power for the instruments on some space satellites. Although they are useful in remote areas of Earth that lack access to conventional sources of power, they are not yet economically feasible for large-scale generation of electricity. The most expensive part of a photovoltaic cell is the semiconductor. Many cells are made with silicon crystals. Films of amorphous (having no crystalline order) silicon are cheaper, but are also less efficient and tend to degrade. To lower the cost, researchers are working to develop cheaper materials to construct semiconductors. Some photovoltaic cells also contain mirrors or lenses to focus the sunlight onto smaller areas. Scientists are also working on materials that mimic the action of chlorophyll in plants, which converts sunlight into chemical energy. Of course, no matter how efficient and inexpensive photovoltaic cells become, they cannot produce energy at night or in inclement weather. We'll still need back-up sources of energy—or we can all change our lifestyle to that of the proverbial farmer who goes to sleep at dusk and rises at the crack of dawn.

How do plants convert sunlight into chemical energy? (See page 121: Photosynthesis)

What are "crystals" of silicon? (See page 53: Crystals and Crystallography)

CURRENT AFFAIRS

Put in the simplest terms, electricity is the movement of electrons, the negatively charged particles that surround the nuclei of atoms. Put a little less simply, electricity is a flow of charge carried by electrons. And put even less simply, electricity is a flow of charge carried by electrons as they are drawn toward the positively charged end of a conductor.

Just as heat will flow from a region of higher temperature to a region of lower temperature until equilibrium is reached, an electric charge will flow from the end of a conductor that has a higher electric potential to the end with a lower electric potential. Potential refers to the energy required to bring a unit of electric charge from an infinitely distant point to the point where it is needed. If you touch a live wire, for example, the difference in electric potential between the wire and the ground under your feet is very great, and the charge will zip right through you, the conductor.

To keep an electric current flowing, it is necessary to maintain a difference in potential. This is the work of an electric generator. A generator takes mechanical energy—the energy of motion—and converts it into electric current. In its basic form, a generator consists of wires rotating in a magnetic field. The magnetic field causes the negatively charged electrons to separate from the positively charged protons, creating a potential difference. As the wire rotates, it flips from one side of the magnetic field to the other, causing the electrons to flow through the wires, first in one direction and then in the other. We call this alternating current, or AC. A battery, which stores chemical energy that is converted to electrical energy, generates current in one direction only; we call this DC, for direct current.

Like other technologies, electricity has its own set of terms, most of them named after famous people. The unit of potential is called the volt, named after Alessandro Volta, who produced an early version of the battery in 1800. The greater the potential, or higher the voltage, the more electrons can move through a conductor. The rate at which an electric current carries energy is called power. Power is measured in watts or kilowatts (thousands of watts)—named after the Scottish engineer James Watt, one of the inventors of the steam engine. The rate of electric flow is measured in amperes, or amps—named after the French physicist André-Marie Ampère. The three terms are related: one watt is the rate of energy transformation (or power) produced by one amp flowing through a conductor whose ends have a potential difference of one volt.

If a generator converts mechanical energy into electric energy, something must provide the mechanical energy. That something can be as simple as a water wheel, or as complex as a huge hydroelectric dam or nuclear power station. The machine that converts the energy of the fluid into mechanical energy is called a turbine. Turbines require fuel. In the past, we've relied on the so-called nonrenewable fuels such as petroleum and coal (since it takes millions of years to build up new supplies, we consider them nonrenewable). The trend now is toward renewable sources such as solar and wind power, as well as nonrenewable nuclear energy.

The basic concept of an electrical current is fairly simple—a stream of electrons passing through a conductor. What is less simple is supplying the energy to keep enough electrons moving to illuminate our cities, and keep us comfortable and entertained.

How is nuclear energy used to generate electricity? (See page 35: Nuclear Energy)

Some batteries generate energy from chemical reactions. How do plants do the same? (See page 121: Photosynthesis)

NUCLEAR ENERGY

Nuclear energy is the answer to the problem of dwindling resources." "Nuclear energy produces unacceptably dangerous radioactive wastes." "Nuclear energy is environmentally friendly, clean energy." "Nuclear energy is an expensive technology that uses money that could be put to better use elsewhere." Many people have opinions about nuclear energy, often without understanding its basic concepts. The first thing to understand about nuclear energy is that there are two types—just as there are two types of nuclear bombs—fission and fusion. Nuclear energy from fission is commonplace; the outlines of nuclear reactors dot the landscapes of many countries. Nuclear energy from fusion is a much more difficult technology, and scientists have taken only the first steps toward achieving it.

Both fission and fusion, whether in bombs or reactors, are what scientists call thermonuclear reactions. In other words, they involve the nuclei of atoms and very high temperatures. In fission, nuclei are split apart into fragments. With some materials, including certain types of uranium and plutonium, the combined mass of the fragments is less than the mass of the original nucleus, and the difference is released as energy. The fragments then go on to bombard other nuclei, causing a chain reaction. In the core of a nuclear reactor—as opposed to a bomb—the chain reaction is controlled. The released energy heats a layer of water that surrounds the core. The water then turns to steam, which runs an electric generator.

The two main dangers of nuclear reactors are the risk of a meltdown, which could leak radioactive material into the surrounding atmosphere, and the problem of how to dispose of the waste products, called radioactive waste. Radioactive waste contains materials with a half-life of thousands of years. A half-life is the amount of time it takes for half of a given amount of radioactive material to decay into nonradioactive material. Radioactive waste is usually buried in canisters deep in the Earth, in geologically stable regions, or below the seabed, away from the edges of the tectonic plates.

In nuclear fusion, which is the opposite of fission, nuclei are fused together. If the combined mass of the nuclei is less than that of the separate nuclei, the difference is released as energy. This is the same process that makes the stars shine. Though scientists have managed to produce a nuclear weapon through fusion—the hydrogen bomb—they are still far from harnessing fusion energy for peaceful purposes. In 1989, two scientists at the University of Utah made front page news when they claimed they had produced cold fusion—fusion that didn't require the extremely high temperatures required by thermonuclear reactions. Though hope springs eternal in the human breast, no one else was able to reproduce their results, and the scientists faded from the media.

One of the attractions of a fusion reactor is that it would produce no radioactive wastes, with their attendant disposal problems. An endlessly renewable source of energy with no dangerous waste products to dispose of may sound too good to be true, and for now it is. But despite the technical difficulties, many scientists think that sometime in the next century, the same process that lights the stars will light our houses and run our fax machines.

How does an electric generator convert steam to electricity? (See page 33: Current Affairs)

Do any stars lose the ability to produce nuclear reactions? (See page 15: Black Holes)

THE ELECTROMAGNETIC SPECTRUM

When you go down to the kitchen at night for a snack, you may not realize that the electricity that powers your refrigerator and the Elvis magnet that holds your shopping list to the refrigerator door are two aspects of a single phenomenon called electromagnetism. An object has an electric charge when its molecules have an excess of negatively charged electrons or positively charged protons. Objects with the same charge repel each other, and objects with opposite charges attract each other. Every magnetic object, whether an atom, a piece of iron, or Earth itself, has two poles. We call Earth's magnetic poles north and south. When an object is magnetized, its atoms line up with their ends pointing to the two poles. As with electricity, like ends repel, and opposite ends attract. Though an object can carry a single—negative or positive—charge, there is no such thing as a magnet with only one pole. A magnet cut in two becomes two smaller magnets, each with two poles.

Whenever an electric charge moves, it creates a magnetic field, and whenever a magnetic field changes, it creates an electric field. Like many things that sound abstract, this relationship was discovered by a scientist attentive enough to learn nature's secrets from the behavior of common objects. In 1820, a Danish physicist named Hans Christian Oersted noticed that switching on an electric current caused a nearby magnetic needle to move.

Both electric and magnetic energy consist of tiny massless particles called photons. Depending on the behavior we are looking at, we say that photons are particles or waves; that is, that they exhibit either particlelike or wavelike behavior. In studying large-scale effects of electromagnetism, scientists usually emphasize their wavelike characteristics. Because every electric field creates a magnetic field and vice versa, there is a constant oscillation between them. This oscillation moves through space in the form of waves that we call electromagnetic radiation. The size and energy level of the waves determine the specific type of radiation. Going from the longest-waved, lowest-energy radiation to the shortest-waved, highest-energy radiation, scientists divide the electromagnetic spectrum into radio waves, microwaves, infrared radiation, visible light, ultraviolet radiation, X rays, and gamma rays.

All forms of electromagnetic radiation travel at the same speed—approximately 186,000 miles per second—which we commonly call the speed of light. Wavelength is the distance between the crests (or troughs) of two waves, and frequency, or energy, is the number of crests (or troughs) per unit of time. In general, the longer the wavelength and lower the energy, the safer the electromagnetic radiation. Radio waves, which have the longest wavelengths and lowest energy, are the safest. Gamma rays, with the shortest wavelengths and highest energy, are the most dangerous.

Radio waves, being harmless and easily detected, are the ideal medium for communication. Microwaves also play an important role in communications. Because they can be more narrowly focused than radio waves, microwaves are easily beamed up to satellites and back to Earth. Long-distance telephone calls and many television signals are transmitted this way (all those TV dishes in people's backyards are microwave receivers). We experience infrared waves basically as heat, though some animals, unlike us, have infrared vision. Visible light, which is central to our daily lives, occupies only a small portion of the electromagnetic spectrum. Though most ultraviolet rays are absorbed by Earth's atmosphere, those that do reach us cause our skin to tan and burn, and can also cause skin cancer. X rays, which penetrate some materials, are especially useful in medicine, in both imaging and treatment. Gamma rays, which are often produced by stars, are absorbed by Earth's atmosphere. Astronomers are able to use orbiting gamma-ray detectors to study them.

Back to the role of electromagnetism in your late-night snack: if you turn on the radio and heat up leftover casserole in the microwave oven, you're making use of radio waves and microwaves. The heat coming from your casserole as you carry it to the table is infrared radiation, and unless you have a yen for eating in the dark, you probably turned on a switch to produce visible light. And if your kitchen has fluorescent lights, they transformed ultraviolet radiation to visible light.

Can any animals "see" infrared radiation? (See page 158: Infrared Radiation)

Can any animals "see" ultraviolet radiation? (See page 160: Ultraviolet Radiation)

Could extraterrestrial civilizations be trying to use radio to contact us? (See page 13: Extraterrestrial Intelligence)

How do we see colors, which are subdivisions of the visible light spectrum? (See page 154: Color Vision)

THE DOPPLER EFFECT

The usual example of the Doppler effect is that of the moving train. Let us say that you live across the street from the train tracks and have just sat down to breakfast. The 8:00 A.M. train is right on schedule. As it approaches, its pitch seems to rise, and as it departs, the pitch lowers. It doesn't matter who is doing the moving. Let's say that you stick your head out the window and yell, "Quiet! I'm trying to read the paper." If the conductor's window is open (and you yell very loudly), the pitch of your voice will seem to rise as the train approaches you, and lower as the train departs. The pitch doesn't really change; the reason it seems to is that as the sound approaches, more waves per second reach the ear, and as the sound recedes, the number of waves per second decreases again. Since pitch is related to frequency—the number of waves per second—an apparent change in frequency results in an apparent change in pitch. The effect, which was predicted in 1842 by a man named Christian Johann Doppler, was actually confirmed by another scientist who conducted experiments on a moving train.

The Doppler effect applies to light waves as well as to sound waves. In fact, Doppler first came up with his idea in an unsuccessful attempt to explain the colors of stars. However, the Doppler effect did become the basis for the concept of redshift. Redshift, also called Doppler redshift, refers to the shifting of the light of certain galaxies toward longer wavelengths, or toward the red end of visible light. Astronomers now know that redshift can occur with radiation from any part of the electromagnetic spectrum; therefore, radiation from the portions of the electromagnetic spectrum that already have longer wavelengths than visible light actually shift away from, not toward, red.

Where, you may ask, are the moving trains in space that cause redshift? The galaxies themselves are the moving trains. The redshift is caused by the fact that these galaxies are moving away from us. The astronomer Edwin Hubble, after whom the Hubble Telescope is named, used this phenomenon to devise what astronomers call the Hubble constant, which is the rate at which the velocity of a receding galaxy increases with distance. In other words, the farther away a galaxy is, the more quickly it is receding. Astronomers use the Hubble constant to calculate how far away a galaxy is. They consider the fact that galaxies are receding to be evidence that the universe is expanding—and has been since the big bang.

You may not travel nearly as fast as one of those receding galaxies, but someone—not an astronomer, but a police officer—may be using the Doppler effect to calculate your speed. Some systems of radar (short for radio detection and ranging) measure speed by measuring the difference in frequency between the radio waves they send out and the waves that are reflected back from the moving vehicle.

Where are these receding galaxies coming from? (See page 1; The Biggest Bang)

WHY THINGS DON'T FLY APART

Ever wonder why you, the chair you're sitting on, and the apple you're eating don't just fly apart? What holds them together also explains many friendships and marriages: the fact that opposites attract. Of course, here we're not talking about short and tall, or introverted and extroverted, but about positive and negative electrical charges.

To find out why things don't fly apart, we have to go all the way down to the atomic level. All objects, including readers, chairs, and apples, are made up of atoms. Atoms consist of protons, neutrons, and electrons. The simplest atom, the hydrogen atom, has one proton and one electron. Many of us remember illustrations of atoms that looked like solar systems, with electrons orbiting the nuclei (protons and neutrons together) like little planets about a sun. This picture is not really true, but sometimes it's convenient to think of it this way. Things don't fly apart because atoms often like to stick to each other to form what we call chemical bonds. Every nucleus has a positive electrical charge (*proton*, positive; and *neutron*, neutral) and every electron has a negative charge. All atoms can exist in a neutral state or in a positively or negatively charged state. When the charges between the nucleus and electrons balance out, the atom is neutral. When there are more protons than electrons, the atom has a positive charge, and when there are more electrons than protons, the atom has a negative charge.

A simple chemical bond consists of two nuclei and some electrons between them. The attraction between the positive nuclei and the negative electrons holds the atoms together. The two main types of chemical bonds are ionic bonds and covalent bonds. In ionic bonding, one of the atoms is much more greedy for the electrons. The normally neutral atom that has extra electrons becomes negatively charged, while the normally neutral atom that loses electrons becomes positive. A charged atom is called an ion. The paired ions are attracted because of their differing charges, but their positive nuclei repel each other. This balance keeps the atoms from getting *too* close together—you might say that when a pair spend too much time together, they start to feel a need for their own space.

In a covalent bond, the electrons are shared more equally by the atoms. As in the ionic bond, the repulsive force between the positive nuclei keeps the atoms from getting too close together. The most important example of covalent bonding involves carbon. In most of the substances it forms, the carbon atom is joined to the other atoms by four covalent bonds.

Many substances consist of both ionic and covalent bonds. The process by which bonds are formed and broken—by which one substance is changed into another—is called a chemical reaction. The beauty of the system is that out of a limited number of atoms—the 109 elements of the famous periodic table of the elements—are created millions of different kinds of substances. Even better, the same two kinds of atoms combined in different proportions can be used to create things with completely different physical characteristics.

So you can say that electricity—or electrical phenomena—holds you together. Technically, the force that holds you together is electromagnetic, but the "electro" part accounts for most of it. Other forces, such as gravity, may keep you from rising up from your chair, but they have little to do with keeping your atoms from spraying around the room.

Is there a limit to how many different substances can be formed from the elements in the periodic table? (See page 49: Making New Elements)

DIAMOND AND GRAPHITE

What do bubble gum, the planet Mars, cellophane, and Limburger cheese have in common? They are all made up of atoms that consist of protons, neutrons, and electrons. Diamond and graphite have even more in common. They are two pure forms of the element carbon. Diamond and graphite are so different from each other because their carbon atoms are organized in different ways. As far as we know, diamond is the most stable arrangement in space of carbon atoms.

Carbon not only makes expensive rings and inexpensive pencils possible; it makes life possible. Chemists call the chemistry of carbon the chemistry of life; they also call it organic chemistry because of its supreme importance to all organisms. Among both living and non-living substances, carbon forms more compounds than all the other elements combined. The typical carbon atom contains a nucleus of six protons and six neutrons, surrounded by six electrons.

Two materials could hardly be more different than graphite and diamond. Graphite is dark, opaque, soft, and smeary; that's why it's used as the "lead" in pencils. It also makes a good lubricant. Diamond is transparent and very hard. In fact, it is one of the hardest substances known. This makes it as popular with industrialists as with jewelers, for its use in cutting tools.

Why do some carbon atoms end up as lowly graphite, while others end up as diamond? Scientists believe that diamond forms deep in Earth's mantle, under extremely high pressure and temperatures over 2,600 degrees Fahrenheit. Gas-driven explosions then propel the diamond-bearing rocks rapidly toward Earth's surface. Most diamonds come from Africa. Large quantities of diamonds also exist on the floor of the Atlantic Ocean, off the African coast. But the cost of mining these diamonds would be much higher than the value of the stones themselves.

In the late 1980s, scientists discovered a new (to them) form of carbon molecule, in the shape of a hollow sphere. Because it so resembled the geodesic domes designed by visionary and inventor

Buckminster Fuller, they named this type of molecule a *fullerene*. They then named one of the larger and more exciting fullerenes the *Buckyball*. Buckyballs are of particular interest to chemists because they can host other atoms or molecules in their cavities.

Astronomers believe that some of the carbon in the debris of a supernova forms graphite and some of it forms tiny diamond particles, each of which contains only a few thousand atoms. Some of this diamond dust has been found in meteorites. So perhaps the cliché about stars shining like diamonds against black velvet is not so fanciful, after all.

What holds the carbon atoms together? (See page 43: Why Things Don't Fly Apart)

WATER

Although water is now chic enough to be sold in designer bottles, most people don't realize just how special it is. Our very lives not only depend on it, but are mostly made of it; each of the individual cells in our body contains water. Most of Earth's surface is covered with water, which rises in the form of vapor, to descend again as rain or snow. We describe extremes of weather in terms of too much or too little water—floods and drought. It's no accident that water is so ubiquitous and so crucial, for many qualities make it unique among the millions of substances that surround us.

The formula for water is simple and famous: H_2O—two hydrogen atoms linked by an oxygen atom. The secret to water's ubiquity lies in the hydrogen bond, a relatively weak bond, for water likes to stick to things—almost anything. Water molecules are in the form of what chemists call polar covalent bonds. In a covalent bond, two nuclei that are bonded together share electrons between them. In many covalent bonds, they share the electrons equally. But in a polar covalent bond, the electrons are attracted more strongly to one nucleus than to the other. This gives the molecule a slight negative charge at one end and a slight positive charge at the other end. In a purely ionic bond, the nuclei transfer all their electrons from one to the other rather than share them. The nucleus that gives up its electrons has a positive charge, and the nucleus that gains electrons has a negative charge. The charged ends of the polar water molecules bond easily with most ionically bonded substances, the negative ends attracting positively charged molecules and vice versa. Water also bonds with many other substances with polar covalent bonds. Thus, many substances like to dissolve in water, from the salt in the oceans to the nutrients in your blood.

Because water molecules are polar, or charged, neighboring water molecules are also very attracted to each other—negative ends to positive ends. If water molecules were not so strongly bonded to each other, it would take less energy to pry them apart, and water would boil at a much lower temperature. Most of the water in the oceans would be vaporized.

When it's not acting as a solvent (the substance in which another substance dissolves), water is busy participating in chemical reactions. It is a major actor in the formation of many polymers, the long chains of molecules that make up everything from the proteins in your body to the nylon that sometimes clothes it. In one type of polymerization, condensation polymerization, small water molecules are released as a chain forms. In the reverse process, called hydrolysis, water molecules interfere with the polymer bonds and the chain falls apart.

Perhaps the most interesting of water's quirks is that, unlike all other materials, it expands when it changes from a liquid to a solid. As the temperature falls, the water molecules pack together in tetrahedrons. A tetrahedron is a four-sided shape, each face of which is a triangle. Because the tetrahedral arrangement takes up more space than the randomly arranged molecules in a liquid, the solid form of water—ice—is less dense than the liquid. This makes the ice cubes in your drink float. More important, when the water on lakes and oceans freezes, the ice floats on top. This allows fish and other life to survive under the ice. If ice sank, the oceans would have frozen solid, and life never would have evolved on Earth. That would give a much more profound meaning to "Water, water every where, nor any drop to drink."

Why do atoms like hydrogen and oxygen tend to stick together? (See page 43: Why Things Don't Fly Apart)

How do polymers form? (See page 51: Polymers)

MAKING NEW ELEMENTS

One way we learn about the world around us is by identifying patterns. Small children catch on to the fact that when Mommy and Daddy get dressed up in the evening, they're going to be left with the baby-sitter. Cats learn that when you open a certain kitchen cabinet, goodies are on the way (and soon they're trying to open the cabinet themselves). When studying a foreign language, we look for patterns of grammar and word usage. A large part of a scientist's task is the search for patterns. The reward is not only the satisfaction of finding out how things fit together, but of using a pattern to predict things as yet undiscovered. A beautiful example of such a pattern is the periodic table of the elements.

In the late 1800s, a Russian chemist named Dmitri Mendeleev perceived a fundamental order underlying the material world. At the time, chemists knew about atoms, and they knew that atoms combined to form substances. They were in the process of determining which substances were elements, and which were formed from the combining of elements. They had already confirmed sixty-three elements. Mendeleev had the genius to recognize that if the elements were arranged according to atomic weight, that is, the number of protons in their nuclei, certain physical characteristics repeated periodically. (Hydrogen, the simplest atom, with one proton, has an atomic weight of 1.) While maintaining the elements in order of atomic weight, he arranged them in vertical rows according to their physical characteristics. Here's where the predictive power of a pattern comes in: in creating his table, Mendeleev left spaces where he thought elements should be, even though those elements had yet to be discovered. For each space, he predicted the atomic weight of the element, and its physical characteristics. Within several years, other chemists were finding missing elements that proved him right.

The periodic table contains ninety-two elements that occur naturally on Earth. Number 92 is uranium, with ninety-two protons in its nucleus. Looking at the periodic table, chemists wondered whether you could create a new element, number 93, by adding a proton to uranium. The answer to this question (yes) is the basis of the elements numbered 93 through 109. Scientists create some of these elements by bombarding uranium with neutrons. Upon colliding, a neutron splits into a proton and an electron. The proton becomes part of the new element, and the electron is propelled away from the collision. Though chemists use more complicated methods to create some of the newest elements, the basic concept is the same: bombarding uranium to produce extra protons.

Scientists call the elements beyond uranium on the periodic table the transuranium elements—beyond uranium. Once an organization called the Union of Applied and Pure Chemistry confirms who is responsible for the discovery of an element, the discoverer—these days, usually a group rather than an individual—has the privilege of naming that element. Number 106, seaborgium, is the only element named after a living scientist, Glen Seaborg, who played a part in the discovery of ten of the transuranium elements.

How far can we go? How many protons can we add to uranium to form new elements? One problem is that the higher an element's atomic number, the less stable, and thus the shorter-lived it is. The unstable element falls apart into stable particles. Element 110, for example, might have a life of only a fraction of a billionth of a second. It is possible, however, that because of the arrangement of the electrons, some elements with even higher atomic numbers could be more stable.

Just after the big bang, the universe consisted of a cloud of hydrogen (1 on the periodic table). As hydrogen nuclei collided, they formed helium and lithium (numbers 2 and 3). Gravity pulled these three elements together to form the first stars. In the fire of these stars, and in the spectacular supernova explosions that signal the death of some stars, more and more collisions eventually created all of the naturally occurring elements, all the way up to 92, uranium. Thus, we and the world around us are truly born of the stars.

What keeps the protons and electrons together? (See page 43: Why Things Don't Fly Apart)

The atomic weight of an element is determined by the number of protons in the atom. What are protons made of? (See page 23: Quarks)

POLYMERS

Most of us know about molecules and atoms—molecules, composed of atoms, are the building blocks of everything around us. That's true, as far as it goes. But many materials consist of giant molecules formed by thousands of small molecules joined together in long chains. These materials are called polymers. *Poly* means "many"—as in *polyphony* (music with two or more melodies) or *polygamy* (marriage with two or more mates). Polymer chains are kind of like line dances at a gigantic party. They usually contain repeating structural units, called monomers. Because the chains of a polymeric material have different lengths (though there is an average typical of each material), there are no specific formulas for polymers. Instead, scientists describe the units that make up the chains.

Polymers are everywhere, including in your body. Proteins, which form hair, muscles, tendons, and skin, are polymers; they consist of long chains of amino acids. Enzymes, which act as catalysts for various bodily processes, are proteins. An example is lactase, which helps the body to break down lactose. Insulin is also a protein. How the body makes proteins is of utmost interest to researchers. The synthetic production of insulin, for example, could be of great benefit to those with diabetes.

The process by which a polymer is created is called polymerization. There are basically two kinds of polymerization, condensation polymerization and addition polymerization. They are what their names sound like. When a polymer is formed by condensation polymerization, small molecules of water or other materials are driven off as the polymer forms. When a polymer is formed by addition polymerization, all of the small molecules go into the chains, with no loss of materials. The chains of most condensation polymers contain 5,000–30,000 small molecules, whereas the chains of most addition polymers contain from 20,000 to several million small molecules (that's some dance!).

Most polymer research focuses on ways to improve natural polymers and to create synthetic ones. Rubber is a well-known polymer that exists in natural, improved, and synthetic forms. Rubber consists of long, wiggly chains that lie in random coils when in a resting position. When the rubber is stretched, the chains uncoil into an orderly arrangement. When it is released, the chains return to their coiled disarray. This is what gives rubber its elasticity. Natural rubber comes from trees. In 1839, Charles Goodyear discovered that heating natural rubber with sulfur caused the chains to bond to each other, making a stronger rubber, called vulcanized rubber. Eventually, chemists developed totally synthetic rubbers with even greater elasticity and strength.

Many polymer chemists work on creating synthetic fibers. Fibrous materials, both natural and synthetic, consist of bundles of chains wound around each other. Most synthetic fibers are organic polymers; that is, they contain carbon atoms. (Organic chemistry, the bane of premed students everywhere, is the chemistry of carbon.) Perhaps the most infamous synthetic fiber is polyester. All those leisure outfits that refuse to wrinkle are the gift of polymer chemists. Lest you shun polyester for more "natural" fibers such as cotton, however, keep in mind that the most common material used to make polyester (and acrylic and nylon) is oil. Clothes made of oil do not sound very romantic, but they are "natural."

Do any polymers contain water? (See page 47: Water)

What happens when an elastic polymer, like rubber, gets stretched too far? (See page 55: The Breaking Point)

CRYSTALS AND CRYSTALLOGRAPHY

Some people think of crystals as pretty pieces of faceted glass used to make earrings and pendants. Others think of crystals as small pieces of quartz with supposedly mystical properties. Yet far more of the world than most people realize is composed of crystals—in fact, most materials in their solid state have a crystalline structure. Gases and fluids are relatively disordered. A crystalline material is ordered at the molecular level. Only a few solids, such as glass, are disordered. Some scientists consider glass to be a supercooled liquid— that is, a liquid that is cooled below its freezing point without having reached a highly ordered, solid state. The crystal structure of solids reveals a great deal to chemists about how atoms of different materials are bonded to one another. It also provides scientists from many fields with a powerful tool called crystallography.

Many crystalline materials, such as metals, are not recognizable as such to the naked eye, yet they have an internal orderly arrangement. A crystalline substance consists of tiny three-dimensional geometric shapes arranged in an orderly, repeating fashion. Scientists call the smallest repeating shape the unit cell. In a repeating wallpaper pattern of kittens and baskets, the smallest grouping of kittens and baskets that a printer could isolate to print repeatedly would be the unit cell.

Scientists characterize a crystal by its geometric shape and its symmetry. All crystals exhibit some types of symmetry. Symmetry describes ways in which you can change the position of an object without changing its shape. The human body, for example, has left–right symmetry, but not front–back or up–down symmetry. That is why you can recognize your mirror image even though it is flipped from left to right. You would have much more trouble adjusting your clothes if the mirror flipped your head to the bottom and feet to the top.

Scientists have learned to use two-dimensional images to determine the three-dimensional structures of many crystals. Their technique, called crystallography, is based on the fact that the wavelength (distance from wave peak to wave peak) of X rays is roughly the same as the spaces between atoms of a crystal. When X rays pass through a crystal, the atoms they come in contact with cause them to diverge at different angles. Scientists call this diffraction. A photographic plate placed on the far side of the crystal records the angles at which the rays have been diffracted. When waves of a beam overlap, they double in intensity, or amplify each other. When a peak and a trough overlap, they cancel each other out. Because the X rays diffract at different angles, some waves amplify each other and others cancel each other out. Where they are amplified, they make darker spots on the image, and where they are canceled, they make fainter spots. Crystallographers use both the spacing of the spots and the differences in intensity to determine the three-dimensional structure of the crystal. They use the technique in fields as varied as physics, mineralogy, metallurgy, and biology.

When a snowflake alights on your coat sleeve, you may admire its lacy six-pointed structure. What you don't see is that smaller units of the snowflake have the same crystalline structure. At the molecular level, the world is a place of much greater order than we are aware of in our daily life.

What holds crystals together? (See page 43: Why Things Don't Fly Apart)

Can the same substance be both a crystal and not a crystal? (See page 45: Diamond and Graphite)

THE BREAKING POINT

Regardless of what the ads say, there is no such thing as an unbreakable material. If enough stress is placed on a material, it will break. *Stress* is the engineer's word for the load, or pressure, that is placed on a material, and *strength* is the measure of how large a stress a material can resist. As with people, you can be fooled. A material that looks strong may break quite easily, and a less impressive-looking material may actually be quite strong.

Materials face two kinds of pressure: tension, or pulling; and compression, or pushing. Engineers say that a material that resists pulling has tensile strength and that a material that resists pushing has compressive strength. Although with some materials the stretching or compressing can appear quite dramatic, the movement takes place within the chemical bonds that hold the substance together. The bonds between the atoms can be thought of as little springs. When a material stretches, the bonds between the atoms lengthen, and when a material compresses, the bonds shorten.

Materials break, or fracture, in two basic ways. They can stretch to a point of extreme deformation before giving way, the way a rubber band or piece of soft toffee does; or they can crack, the way a piece of glass or porcelain does. When a tensile, or ductile, material finally gives way, the atoms slide past each other; finally the shearing becomes too extreme, and the material separates. When a brittle material cracks, bonds between layers of atoms at the site of the initial crack break. The pressure that was transmitted uniformly throughout the material now has to go around the crack. This places even greater pressure at the area just beyond the crack, and more bonds break. This causes a chain reaction of breaking bonds, which is why an object can appear to shatter so quickly.

Just as there is no such thing as an unbreakable material, there is no such thing as a totally rigid one. If you are leaning your elbow on the table as you read, the table is giving way, however slightly, under the pressure of your elbow; it is temporarily deformed by the load of your elbow. The ability of a material to return to its original shape when a load is removed is its degree of elasticity. Not only the most obvious materials and objects, such as rubber bands or mattresses, are elastic. The floorboards beneath your feet must return to their original shape after you walk across the room.

A material that does not return to its original shape is said to behave plastically. A familiar example is the elastic waistband that stretches out and loses its shape; it gets slacker and slacker before it finally breaks (if you haven't already thrown it out). A brittle material behaves elastically until it reaches its breaking point, and then it breaks suddenly without warning. That is why some strong materials such as glass can't be used in certain large structures: they would give no warning before they failed. You might say that an elastic material is like a person who easily gets "bent out of shape," and that a brittle material is like a person who always acts as if everything's okay and then one day suddenly falls apart.

Some people confuse stiffness with strength. A stiff material can be strong or weak, as can a flexible material. Let's say you're having a midnight snack while reading this book. The grape jelly you're eating is flexible and weak—see how easily pieces break off and plop on the table. The crackers are stiff, or brittle, and weak—they break only too easily under the pressure of the jelly knife. The plate under the crackers is stiff and strong, and the dental floss that you'll be sure to use before going to bed is flexible and strong.

How does chemical bonding within atoms and molecules occur? Why does it make things hold together until forced past their breaking point? (See page 43: Why Things Don't Fly Apart)

Can scientists design substances that have different chemical compositions for different kinds of responses to stress? (See page 51: Polymers)

SOAP ✓

Soap's great abilities, you might say, lie in its split personality. Not because it comes in pretty wrappings but works like a horse, not because it can seem so gentle and then work itself into a lather, but because its very molecules can't decide what they want.

All soap molecules contain a long hydrocarbon tail (a hydrocarbon is a compound that consists just of carbon and hydrogen) and a water-soluble head. Herein lies the soap molecule's confused self-identity. The head of the soap molecule is highly polarized and thus attracted to polarized substances such as water. It is called hydrophilic, which means water-attracting. The other end of the soap molecule, the tail, is nonpolarized and attracted to nonpolar substances such as oils. It is called hydrophobic, which means water-repelling.

When soap dissolves in water, it forms large clusters of molecules called micelles. The hydrophobic tails face in toward the center of the micelles, and the hydrophilic heads face out. When these soap micelles, which are in constant motion, meet a greasy object such as your face, the tails attract the grease, while the heads form bonds with the water. In effect, the soap allows the grease to act as if it were water-soluble. Because the soap acts at the surface where the greasy object and water meet, it is called a surface-active agent, which is usually shortened to *surfactant.* Detergents, which are synthetic substances that act as soaps, are also surfactants.

The micelle, with its inner tails buried in grease and its outer heads bonded to the water, forms a grease droplet. The grease droplets, which are prevented from gobbing together by the repulsion of the charged heads of the soap molecules, are then carried off by the water. The face or shirt is now clean. Vigorous agitation at the interface between the soapy water and the greasy object greatly aids the formation and removal of grease droplets. In other words—scrub hard!

All soaps consist of fats and alkalis, or soluble salts. Soap was probably discovered by early ancestors cooking over a fire, when grease from a meal fell onto the wood ashes. If water was poured onto the ashes to douse the fire, the grease and the lye that was leached from the ashes made a primitive kind of soap. Some person on kitchen—or campfire—duty noticed that this "soap" had the useful ability to get things clean.

Today, soap is usually made with palm or olive oil instead of beef tallow, and with sodium hydroxide instead of lye. Different formulas make soaps that lather more or less easily. When soap is used with hard water—water that contains calcium and magnesium ions from rocks near the water source—it forms an insoluble material called lime soap. This greasy deposit is also known as a bathtub ring—something our early ancestors didn't have to worry about.

Why would there be carbon in the ashes left over from a fire? (See page 62: Fire)

THE CHEMISTRY OF THE BRAIN

The entire human body is made up of cells, and all bodily growth and activities are the result of chemical changes within those cells. But what about what's inside our head? Not just our brain, but everything we associate with it—our thoughts, our emotions, our memories, our consciousness, our very sense of self? To some scientists, all brain activity can be reduced to chemical processes. Others go only so far as to say that although chemistry explains a lot, it's premature to say whether it explains everything.

The billions of specialized nerve cells of the nervous system are called neurons. Neurons, the largest of the body's cells, are distinguished from other bodily cells and tissues by the elaborate system of communication between them. It is through this communication network that the brain sends and receives messages to and from other parts of the body—as well as ponders the state of the world. Sensory neurons receive information about sensations such as touch and temperature and convey it to the brain. Motor neurons send messages from the brain to the periphery, to activate muscles and glands. Interneurons, the most numerous, are interconnected with sensory neurons, motor neurons, and other interneurons; they organize complex activities of the body and brain.

Each neuron has a center, called a cytoplasm, surrounded by a membrane that contains proteins and other chemicals. The membrane acts as a barrier between the cytoplasm, which has a negative charge, and the fluid outside the cells, which has a positive charge. This difference in charges is of great importance to how impulses travel from one neuron to another. At one end, the neuron has branched extensions called dendrites, which receive incoming impulses and pass them to the cell. At the other end, the neuron has a long, threadlike extension called an axon, which passes impulses to other cells, sometimes over great distances. The axon ends at what is called—appropriately enough—a nerve ending. The axon of each neuron usually sends impulses to many other neurons, and the dendrites receive impulses from many other neurons.

When an impulse travels from the axon of one neuron to the dendrites of another, it jumps over a tiny gap called a synapse. Each neuron is a one-way street; that is, impulses go in only one direction. The impulses are transmitted by a chemical process that causes an electrical one; thus, the signals are called electrochemical signals. The process begins when an axon releases a chemical called a neurotransmitter across a synapse to the dendrites of another neuron. When the neurotransmitter binds to the receptor cell, the proteins of the receptor cell "open" in a way that allows positively charged particles to pass through. These positively charged particles, which are attracted to the negative charge of the cell's interior, change the balance so that the interior of the cell becomes positively charged. Nearby proteins in the cell then open in similar fashion, allowing positively charged particles to enter, and then close again. As each region of the cell reverses its charge, the preceding region reverts back to the normal, negative charge. This process, which continues along the length of the cell, is what we call a nerve impulse.

In addition to the neurotransmitters that instruct neurons to fire impulses, there are other neurotransmitters that inhibit neurons from firing. Without this balance, all our neurons would be constantly transmitting impulses, which would surely be an uncomfortable—if not unbearable—state.

Perhaps the problem in thinking of our nervous system as mere chemistry is the word *mere*. Our nervous system is like an exquisitely complex orchestra of tiny instruments sending electrochemical messages. The mixture of messages is incredibly nuanced and endlessly changing. If there is a conductor, the conductor is our self. There is nothing "mere" about it.

Are the objects of our sense impressions really "outside" us, or are they all in our heads? (See page 151: The Senses)

What is the electrical basis for the brain's activity? (See page 33: Current Affairs)

What role does brain chemistry play in memory? (See page 195: Memory)

Does "thinking" require a brain? (See page 212: Artificial Intelligence)

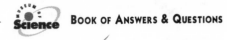
FIRE ✓

For an example of the conversion of matter into energy, you can't do much better than fire. Think of that pile of logs next to the fireplace; imagine the beautiful flames and cozy warmth, and later, the small pile of ashes. Think of the charcoal carefully arranged under the backyard grill, the flame that cooks your steak or swordfish— and then of the black residue that remains. Even the paper that you use for kindling diminishes dramatically.

Fire is a chemical reaction in which a material combines with oxygen, or is oxidized. The oxygen can be part of an oxygen-containing compound. The process of burning, which is called combustion, begins when the fuel is heated enough to cause the jolted molecules to interact with the available oxygen. Matches are designed to ignite from the low level of heat caused by the friction of the match being rubbed against another surface. The substances that result from combustion have less energy than the original substances—the fuel and oxygen—and the excess energy is released as heat and light. A small amount also takes the form of waves of air compression; this is the sound you hear as something burns. Though the basic concept is simple, the actual process is complex. In the simplest combustion reaction, that of hydrogen and oxygen, the atoms go through at least fourteen steps of grouping and regrouping.

Fuels can be solid, liquid, or gas. When a solid, such as wood, combusts, matter that vaporizes at relatively low temperatures is forced out and burns first. This is why it is sometimes hard to get a wood or charcoal fire going. With liquid fuels, the heat evaporates liquid at the surface of the fuel, and it is this vapor that burns. As the vapor is burned off, more vapor evaporates. As the temperature of a gas is raised, high-energy collisions between the molecules cause the bonds between the atoms to break. The atoms then react even more rapidly, and combustion occurs as a chain reaction. If temperatures are high enough, the chain reactions take place so rapidly that the entire fuel supply combusts simultaneously, or explodes. In other words, an explosion is just a very fast fire. Regardless of whether a fuel is solid, liquid, or gas, a fire will burn as long as there is fuel and a sufficient supply of oxygen. When you smother a fire by throwing a blanket over it, you deprive it of oxygen.

Spontaneous combustion is not magic and it does occur, usually when materials are stored in bulk. The temperature of the material rises as microbes in it produce heat. Because there is no place for the heat to escape, the temperature rises even higher. This continues until finally the material bursts into flame.

We started out talking about the conversion of matter into energy. If we take a step back, we see that the fuels we use were formed from the conversion of energy into matter. The tree that provides our wood soaked up the energy of the Sun and used it to grow. Fossil fuels such as petroleum release solar energy that was stored millions of years ago. Now that's recycling for you!

You can feel the heat from a fire, but can any animals actually see that heat? (See page 158: Infrared Radiation)

ZERO ✓

The concept of zero in the Western Hemisphere may have been invented sometime in the first few centuries C.E., at a Mayan cocktail party as a waiter made the rounds with a tray of oysters on the half-shell, for the Mayan symbol for zero is an empty oyster shell. Most historians, however, focus solely on zero's invention by the Babylonians, in about 500 B.C.E., and its reappearance in India about one thousand years later. From there, zero traveled to the Middle East, where it joined the Arabic number system and then continued on to Europe and the Americas.

This well-traveled zero is zero as a position holder, the zero that indicates the difference between 2, 20, and 200. This zero tells how much value to place on the other digits. The zero as place holder grew out of the practice of counting with beads. In many early cultures, people counted with beads arranged in columns on a counting board. The columns functioned the same way the columns in our notation do. A bead in the first column had the value of one, a bead in the second column had the value of ten, and so forth. Before the invention of zero, however, the place where we put a zero was empty. A ten was indicated by a bead in the second column, next to an empty space in the first column. We are so used to using zero that it is hard for us to conceive how earlier people could have had a positional system so similar to ours without coming up with a symbol for the empty position.

After the Babylonian invention of zero as a place holder, it took another thousand years for zero as a number to be discovered, in the sixth century, by the Hindus and Chinese. It then took another seven hundred years before it came into use in the Western world. Apparently the concept of zero as a number was so foreign to the human mind that even negative numbers were discovered earlier.

Does zero as a number refer to anything more—perhaps we should say other—than nothing? Yes and no. Zero does refer to nothing, but it is a unique number with specific mathematical properties. The whole numbers, 1, 2, 3, 4, 5, etc., are called integers. Zero, too, is an integer. It is the only integer that is neither positive nor negative. The integers fall in consecutive order. One way to demonstrate that zero is an integer is to count from –5 to +5. You must place a zero between –1 and +1, for they are not consecutive.

A number is an abstraction. The number 2 can stand for two people, two elephants, or two candy bars. There is no limit to how many pairs of things can be designated by the number 2. Mathematicians say that the number 2 refers to all those sets that contain a pair of things. The number 3 refers to all those sets that contain three of anything, and so forth. Since there is no limit to how many such sets there might be, mathematicians say that the sets are infinite. Zero is the set that contains no representatives of any item; thus mathematicians call it the empty set.

Though we usually begin counting items with the number 1, to be mathematically correct we should begin with zero. Say we are counting tangerines. To the empty set of no tangerines, we add one tangerine to reach the set of one. We then add a second tangerine to reach the set of two, and so forth. Of course, we could change our terms and add a tangerine, cantaloupe, and watermelon to the empty set of no fruit, to reach the set of one fruit salad.

What do scientists mean when they say that number sets are infinite? (See page 83: Infinity)

IRRATIONAL NUMBERS ✓

A child learns to count using his or her fingers: one, two, three, etc. Mathematicians call these whole numbers integers. The next step is fractions: you get half of the last slice of chocolate cake; your little sister gets the other half. Though kids fighting over left-over cake don't usually realize it, a fraction is a ratio of two numbers. In the case of 1/2, it's the ratio of 1 to 2. Until the time of the early Greeks, mathematicians thought that all numbers could be expressed as ratios, or fractions (a whole number can be written as the number over 1, e.g., 4 can be written as 4/1). Then in the sixth century B.C.E., Pythagoras wrote his soon-to-be-famous Pythagorean theorem that the square of the hypotenuse of a right triangle is equal to the sum of the squares of the other two sides. But if each side of a right triangle is one unit in length, then the diagonal must be $\sqrt{2}$. Pythagoras was dismayed to discover that $\sqrt{2}$ could not be expressed as a fraction. In his exasperation, he called $\sqrt{2}$ irrational— and thus was born the concept of irrational numbers. Historical rumor has it that Pythagoras' followers were so upset that they tried to keep the knowledge of irrational numbers from spreading beyond their group.

If the idea of ratios is a little too abstract for you, think instead of units of length. If you take a right triangle whose sides measure 1 inch and 2 inches, then the hypotenuse must equal the square root of $1^2(1) + 2^2(4)$ or (sqr(5)). But you cannot express (root5) in whole inches. You'll always end up with a leftover fraction of an inch.

There's a simple way to determine whether a number is rational or irrational. When you express a rational number as a decimal, you find a repeating sequence. For example, 4 1/3 is 4.3333. . . . An irrational number never repeats. Perhaps the most famous nonrepeating decimal is π (pi), which is 3.14159265358979. . . .

Once people got used to the idea of irrational numbers, things got even worse. It turns out that there are many more irrational numbers than there are rational ones! But there is comfort to be taken in irrational numbers—they fill up the spaces between fractions. Whole numbers and fractions are discrete, or separate, units. No matter how small the fraction, you can always divide it into even smaller fractions. With only whole numbers and fractions, we could not describe a continuous line mathematically. Irrational numbers allow us to do so by filling the spaces between fractions.

Though the world of irrational numbers is much vaster than that of rational numbers, most of us count, measure, and calculate our daily lives in rational numbers. We leave the irrational numbers to the mathematicians, who find a world of numbers that refuse to be captured in fractions fulfilling rather than frustrating.

What is pi? (See page 69: Pi)

If an irrational number never repeats or ends, is it infinite? Is there an infinite number of irrational numbers? What about rational numbers? (See page 83: Infinity)

PI ✓

The quantity π (pi) has a long history, wears many mathematical hats, and could compile its own *Ripley's Believe It or Not* of heroic attempts to calculate it. This weird, inexactly known number is enormously important throughout the mathematical description of our physical world. π—3.1415926535—is the ratio of a circle's circumference to its diameter ($\pi = C/D$). No matter what numbers you put in the C/D side of the equation, π will never be expressible as a ratio of two whole numbers; thus it is what mathematicians call an irrational number. It is also what mathematicians call a transcendental number; that is, it is never the solution to an algebraic equation. (Algebra is a branch of mathematics that uses symbols to stand for different quantities and solves for the value of an unknown quantity by manipulating the symbols on both sides of an equation according to strict rules.)

More than two thousand years ago the Greeks pondered π. Archimedes (ca. 287–212 B.C.E.) used the idea of the polygon to approximate π (a polygon is a figure with a number of sides, e.g., a pentagon or hexagon). He fit a circle between an inner, or inscribed, polygon and an outer, or circumscribed, polygon, and π fell between the limits determined by the two polygons. Using this method, Archimedes achieved an approximation for π of 3.1418. Although he couldn't prove that this number, which falls between 223/71 and 22/7, was an irrational number, he knew that it was an approximation. In the eighteenth century, mathematicians showed that π was an irrational number, and soon after they showed that it was also a transcendental number. The consequence of π's being transcendental is that it is impossible to construct a circle whose area is the same as a given square. We still use the term *circle-squarer*, coined by the Greeks, to mean a person who attempts the impossible.

It is not surprising that π is used in formulas that measure the areas of figures such as spheres and cylinders; after all, they are round. What is surprising, is that π is also involved in the measurement of the area of every closed figure (a *U* is an open figure;

an *O* is a closed figure). Today, π plays a role in many areas of mathematics far more complex than measuring the area of a circle. It is also critical to the statement of the laws of electromagnetism. It even has practical applications in this age of the computer. Computers use algorithms to calculate π (an algorithm is a method of solving a problem through a series of steps). New algorithms developed to calculate π can have other computer applications. Also, calculating π is a great way to work out bugs in computer hardware and software.

Another characteristic of irrational numbers besides their refusal to be expressed as a ratio, is that when calculated as a decimal, they show no repeating pattern. The attempt to calculate π has entered some people's lives as a pastime, and others' as an obsession. In the late sixteenth century, a German mathematician named Ludolph van Ceulen calculated π to 35 places. When he died, his value for π was engraved on his tombstone. Germans took to referring to π as the Ludolphian number. Over the centuries, the π-obsessed calculated π to tens and then hundreds of places. Now the age of the computer has taken the challenge to entirely new levels. In the early 1990s, two Russian brothers living in New York, David V. and Gregory V. Chudnovsky, used a computer they built themselves in their apartment to calculate π to 2.16 billion decimals. The Chudnovsky are not kooky hobbyists; they are researchers at Columbia University. Their hope is to find a subtle pattern that has thus far eluded mathematicians. But if there is such a pattern, it will take more than a couple of billion digits to reveal it.

What is an irrational number? (See page 67: Irrational Numbers)

THE MÖBIUS STRIP: THE ONE-SIDED SURFACE

The Möbius strip is a serious mathematical discovery that moonlights as a parlor trick. Take a strip of paper, twist one end 180 degrees, and tape the two ends together. Then take a pen and start drawing a line anywhere on the strip. Keep drawing until you reach the beginning of your line. Does the line cover both "sides" of the paper? Does this mean that the strip has only one side?

The Möbius strip does indeed have only one side (and one edge), but that is not what mathematicians find most interesting or significant about it. It is not how mathematicians think about Möbius strips. They speak of the Möbius strip's nonorientability. To get a feel for what is meant by nonorientability, make another Möbius strip. This time take your pen and start drawing capital *E*s, the kind you see on eye charts. As you continue to draw *E*s, eventually you will notice that the *E*s on the part of the strip behind the part where you are drawing are facing left instead of right. Thus, you cannot say that the *E*s on your Möbius strip face either left or right.

A common way to fix orientation is with the two directions clockwise and counterclockwise. Make another Möbius strip, and this time instead of *E*s draw small circles. Inside each circle draw a curved arrow going in a clockwise direction. If you hold the strip up to the light, you will see that the arrows on the other side of the paper are going counterclockwise. But there is no "other side" of the paper, because we already showed that the Möbius strip has only one side!

Mathematicians also say that a surface is orientable if it has two sides and you can paint one side, say, black and the other, say, white, and the black and white will not meet except at the boundary of the two sides (the edge). If before twisting your Möbius strip you paint one side black and the other white, when you tape the ends together the black and white will meet at the seam.

Möbius strips can be formed with more than one twist. Take a long strip of paper and give it three twists before taping the ends together. As you can show by drawing another line with a pen, you have again made a Möbius strip. Now take a scissors and cut along the line you just drew. The result is no longer a Möbius strip (as you can show by drawing a line in pencil or another color ink), but a type of knot called a trefoil knot. This way of forming a trefoil knot actually has a real-life application—at least for chemists. Like people, and Möbius strips, molecules have handedness; that is, they are either right-handed or left-handed. Chemists are very interested in how you can change the physical characteristics of a substance by changing the handedness of its molecules. One way they manipulate handedness is by synthesizing, or making, knotted molecules that they can knot in opposite directions. So what does this have to do with Möbius strips? Well, chemists have discovered that one way to form a knotted molecule is by the same method you just used with paper. Basically (though, of course it's much more complicated than this), they make a Möbius strip with the molecule, then cut the strip to get a knotted molecule.

The person who discovered the Möbius strip, a German astronomer and mathematician named August Ferdinand Möbius, is considered the father of the field of mathematics called topology. Topology describes geometric shapes that are unchanged by such deformations as stretching or twisting. A football, for, example, is topologically the same shape as a baseball; that is, a football is like a squeezed baseball. Another way to describe topology is to say that it is the mathematics of continuity—and the Möbius strip is the clearest example of a continuous surface.

If chemists are interested in making new molecules, what kinds of things can be done with these new molecules? (See page 51: Polymers)

When a Möbius strip is cut, it results in a "knot." What, then, is a knot? (See page 73: Knot Theory)

KNOT THEORY ✓

Our first exposure to knots is usually when we learn to tie our shoes. Some of us stop there, or even regress to Velcro; others may advance to sailing knots or macramé. For a growing number of scientists, however, the study of knots is their life work. Among them are mathematicians, physicists, chemists, and biologists.

To make the kind of knot mathematicians study, you tie a knot and then glue the two loose ends together. The basic question in knot theory, which is trickier than it sounds, is how to tell one knot from another. Basically, two knots are the same if by pushing and pulling—but not cutting the loop—you can change one knot to the other. Failing to do so, however, does not prove that the knots are different; you just may not have tried the correct yank or twist.

Mathematicians classify knots by the number of crossings, places where the string crosses over itself. They call the circle, which has no crossings, an unknot. The simplest knot is the trefoil, a circle that winds through itself. The trefoil has three crossings. Mathematicians have identified almost thirteen thousand knots that have thirteen or fewer crossings. But though it's fun (for mathematicians, at least) to catalogue knots, the real excitement lies in the application of knot theory to other fields of mathematics and science.

If each knot had a different label, it would be easy to distinguish one knot from another. In the early 1980s, a Berkeley expert in algebra named Vaughan Jones devised a type of algebraic expression, called a polynomial invariant, that could be used to label knots. It is easy to tell one polynomial from another, so mathematicians can use the polynomials to tell different knots apart. Though these polynomials appear to capture the essence of knots, mathematicians don't yet understand what that essence is. And though two knots with different polynomials are definitely different, the polynomials cannot distinguish between all types of knots.

Until recently, scientists studying DNA concentrated on the arrangement of the base pairs that make up our genetic code, as if the DNA strands were straight lines. But DNA strands form knots and links during replication and recombination (gene-splicing), and then straighten out during cell division. Some biologists are now looking at how the knots are made and unmade. In gene-splicing, biologists use particular enzymes to split apart a strand of DNA and then reinsert the segments elsewhere. By using the polynomials that Jones developed, they can determine the sequence of knots that occurs in the process; they even predicted a knotted form they hadn't seen yet.

Chemists are interested in how different arrangements of the same atoms result in molecules with different characteristics. Knotting a single molecule is more difficult than knotting a strand of DNA. Unless the molecule has many atoms, it is not very flexible. It is also harder to manipulate a chain of atoms. Chemists have developed several methods of synthesizing knotted molecules. One method involves using a metal template to hold the atoms in place while the knot is formed and then removing the template. Another method involves twisting a molecule into a Möbius strip and then snipping apart the Möbius strip to form a knot.

One of the most abstract applications of knot theory is in the field of particle physics. In some cases, the same mathematics that describes a knot crossing can also be used to describe the interaction of two particles. Eventually, physicists trying to decide whether two theories describing the forces between particles are truly different may be able to convert their problem into a question about knots and knot invariants.

There may be no end to the possible applications of knot theory. Some physicists even propose that the entire universe consists of knots embedded in the fabric of space-time. So the toddler struggling with a shoelace may actually be a cosmologist-in-training.

What is a Möbius strip? (See page 71: The Möbius Strip: The One-Sided Surface)

Can scientists alter genes, and if so, to what end? (See page 178: Genetic Engineering)

PROBABILITY

Probability theory, which plays an important role in many fields of science and mathematics, had its start in gambling. Long before casinos and state lotteries—perhaps as long as forty thousand years ago—people rolled dice carved from animal bone. The first manual on games of chance was written in the 1500s by an Italian physician and mathematician named Cardano, and in 1718, Galileo published a manual entitled *Sopra le Scoperte de i Dadi* (On Discoveries about Dice). Of course, then as now, an interest in the finer points of playing dice doesn't necessarily imply knowledge of the mathematics of probability. In fact, many people who like to gamble—or just need to decide whether or not to carry an umbrella—subscribe to beliefs that are contrary to the laws of probability.

The most familiar example of the workings of probability is the tossed coin. What is the chance that a tossed coin will land heads up? Tails up? The answer seems ridiculously obvious—50 percent. But how mathematicians arrive at the probability of an event's occurring is similar to how you would determine the probability of heads or tails. If you toss up a coin one hundred times and record how many times it lands heads up and how many times tails up, you *probably* will get a number around fifty for each. But not necessarily. The larger your sample, the more reliable your results. For example, if your sample consists of only ten tosses, you might easily come up with six, seven, even ten heads. With a sample of one hundred tosses, the chance that they will come out all heads or all tails is so low that were it to occur, you would assume the coin had been altered to favor one side. Part of the job of statisticians is to determine how large a sample is necessary for reliable results. In businesses that depend on statistics, such as insurance companies, larger samples cost more and take more time, but the consequences of using too small a sample can be even costlier.

Some aspects of probability theory are counterintuitive—that is, the reality is the opposite of what "feels right." If it hasn't rained in six weeks, you may be more likely to take an umbrella than if it hasn't rained in two weeks, even if the forecast is the same "partial cloudiness, 30 percent chance of rain." If your aunt wins the million-dollar lottery, you may suggest that she stop buying tickets, since she's not likely to win again. Well, you'd be wrong on both counts. Assuming that the weather conditions are the same, it's no more likely to rain after six weeks of dryness than after two weeks; and your aunt stands the same chance of winning the lottery a second time as she did the first time (very small). The reason for both is the same—they are independent events. In other words, the weather system doesn't know (or care) how long it is since it last rained, and the lottery tickets don't know (or care) who won last week. That is also why you might get seven or eight heads out of ten coin tosses. When the coin is tossed, it has no awareness of the previous tosses. Probability is solely a matter of statistical averages.

Probability theory has given rise to other, related fields such as queuing (waiting in line, in plain language). Banks use queuing theory when they decide how many teller windows to have open at different times of the day. You use queuing theory when you decide whether to stop at the bank during your lunch break or later in the afternoon. Is it better to go at lunch when there's a long line and four windows open, or after lunch when there are few people but only one window open? Of course, you may go when there's one window open, only to find twenty people waiting in line.

Probability enters all realms of science, including quantum theory, which deals with the transformation of energy at the atomic level, and entropy, which is the tendency of order to go toward disorder. But no matter how sophisticated the science, the rules of probability are the same ones that govern tossed coins and lottery tickets.

How does probability relate to entropy? Is it more probable that a system will move from order to disorder, or from disorder to order? (See page 21: Entropy)

Probability theory is one method of making decisions when playing games. What are some other ways of thinking about games? (See page 77: Game Theory)

GAME THEORY

Often when making a decision, we weigh a number of factors such as possible positive and negative outcomes and what moves we think others will make. We base our decisions on thoughts, feelings, and hunches. Let's say, for example, that you're planning a party. Alice and Herman get along with each other, but neither can stand Roscoe, who is a valuable job contact for you. You dare not invite all three. You're tempted to invite Alice and Herman, and not Roscoe; but you're afraid that your cousin Ruby, whom you've already invited, will tell Roscoe about the party. You're in the process—or throes—of what mathematicians call decision-making. Though you may throw up your hands and base your decision on something as arbitrary as the number of hot dogs in a package, mathematicians have devised formal rules for decision-making. Together, these rules are called game theory.

Game theory was founded by John von Neumann, a Hungarian-born mathematician who was on the team that designed the first digital computer. In 1944, von Neumann and an American economist named Oskar Morgenstern published *Theory of Games and Economic Behavior.* Though they based their ideas on the strategies and moves of games, from the beginning they intended their work to help economists, military planners, and others to make rational decisions.

According to game theory, in any game there is an optimum strategy for each player. Each player's strategy has to take into account factors such as random moves and other players' strategies. In the simplest game described by game theory, the two-person zero-sum game, one person's gain is the other's loss. Each player need only try to be the one to gain. The next stage of development involves equilibrium, whereby each player tries to minimize the loss the other player can cause him or her. More complex games involve more than two players and can include coalitions.

The philosophical basis to game theory is that there are two basic ways people can deal with each other: cooperation and exploitation. In many instances, a player's strategy involves choosing between the two. A famous game called the Prisoner's Dilemma highlights this choice. In the Prisoner's Dilemma, there are two prisoners. Each is asked whether the other committed a crime. The dilemma for each prisoner is whether or not to cooperate with the other. Let us say that if they cooperate (each denying the other's guilt), each prisoner gets five points. If both defect, accusing each other of guilt, each gets two points. If one cooperates and the other defects, the one who cooperates gets nothing, and the defector gets seven points. At first glance, it seems best to defect in the hope of getting seven points. But if each prisoner thinks this way, each will get only two points. If more than one game is played, each player also has to consider how his or her decision will affect the other's future decisions. The game has implications for many real-life situations such as trade and diplomacy. Though there are rewards for cooperation, the one who exploits often gains more, at least in the short term. Strictly speaking, of course, game theory ignores the moral and social context. The one who exploits may have to live with a guilty conscience and the disapproval of his or her peers.

How can game players determine what is the most likely outcome of a move or decision? (See page 75: Probability)

Can we make machines that understand game theory, or at least can play games? (See page 212: Artificial Intelligence)

CHAOS ✓

Scientists and philosophers used to think of the world as one huge mechanism, a giant clock of many gears. Astronomical observations supported this view: the Sun rose and sank each day (that is, Earth rotated), the planets orbited the Sun, and the Moon went through its phases. These phenomena could be counted on to continue to occur as they had in the past. Not only did everything fit in place with everything else, but by knowing all the present conditions, one could predict the future. The nineteenth-century astronomer and mathematician Pierre-Simon de Laplace went so far as to claim that if one knew all the initial conditions of every molecule in the universe, one would be able to predict the entire future of the universe. For an example on a scale much more modest than the entire universe, if you line up a cup, a plate, and a saltshaker next to each other, you can accurately predict that if you shove the cup six inches in the direction of the plate, it will push the plate six inches, which in turn will push the saltshaker six inches. But what about more complex systems? What if the cup pushes three plates, and each plate pushes eight saltshakers?

Enter chaos theory. Chaos theory describes the behavior of large systems, such as bodies of flowing water or weather systems. Chaos does not necessarily mean chaotic in the common sense of the word—like the mess under your bed if you toss your clothes and books there every night for a week. Chaotic behavior is unpredictable. The reason it is unpredictable is because it is what scientists call "extremely sensitive to initial conditions." The classic example is that, theoretically, the flapping of a butterfly's wings in the Amazon could cause a tornado in Texas. In other words, the effects of the small movements of air set off by the butterfly could gain in magnitude and complexity until they have built to a tornado in Texas. The sensitivity to initial conditions implies that a slight difference in initial conditions can lead to dramatically different results down the line.

The unpredictability of chaotic behavior is not the same thing as randomness. A random event truly could have gone one way or another. In most instances of what we call random behavior, we are simply ignorant of the factors that caused the result. Even a tossed coin is affected by subtle pressures of the tossing hand and air currents. The only truly random events occur at the atomic level. The precise moment that a radioactive nucleus decays is inherently unpredictable; quantum behavior can be predicted only on a statistical level.

Most chaotic systems show general patterns but are unpredictable in the details. If you sit by the bank of a stream (or, less romantically, on the curb of a gutter after a rainstorm) and watch the water flow by, the ripples made by the water are similar yet constantly changing. Many chaotic systems display a regular pattern for a while, turbulence erupts, and then the system returns to its regular pattern. The stream may flow quietly, interrupted by sudden swirls that unpredictably appear and subside. One reason such turbulence is so difficult to predict is that it takes place on all levels—within a larger swirl are smaller swirls, all the way down to moving drops.

Chaos theory is a scientific and philosophical tool that bridges the gap between order and randomness, and between control and helplessness. Chaos theory gives scientists a tool for studying the behavior of phenomena ranging from irregular heartbeats to star formation.

What accounts for some of the "general patterns" that emerge in chaotic systems? (See page 81: Strange Attractors)

STRANGE ATTRACTORS ✓

Sometimes in mathematics or science, a visual image is more than a means of expression, something deeper than an illustration—it is the very embodiment of a concept. Such is the case with an attractor. An attractor takes form in a special kind of graph that scientists call phase space. In phase space, a scientist plots each of the variables of the mathematical description of a system. A system can be as simple as a swinging pendulum. If the pendulum is swinging in a frictionless environment—and thus is going to swing forever, never to slow down—two dimensions need to be plotted, location and velocity. The resulting graph of these points would describe a circle or a loop. Even if the pendulum wobbles—for example, because of an air current—the path will basically stay within this circle or loop. A pendulum that is subject to friction will eventually wind down to where its location is stationary and its velocity is zero. In this case, the attractor is a single point. In other words, the trajectory of the pendulum is drawn into—attracted to—a single point.

Any scientist (and many schoolchildren) can predict the fate of a swinging pendulum with no external source of power. But many phenomena are unpredictable; they are what scientists call chaotic, or turbulent. One of the best examples is that of flowing water. From the water that you let out of your tap to a roaring river, the whorls and eddies are endlessly varied, with periods of relative calm within the turbulence. Since the path of a turbulent system is unpredictable, by definition it never repeats—for once it repeated, the rest of the pattern would then be predictable. In 1963, a meteorologist named Edward Lorenz plotted the mathematical description of atmospheric air flows, certainly a case of turbulence. Plotting the variables of the air flows resulted in a series of graphs (phase space diagrams) that resembled figure eights, or butterfly wings. Though it was impossible to predict the paths of the air flow exactly, the flows always fell within the butterfly shapes described by these phase space diagrams. As abstract as the image was, it also gave some sense of the physical motion of the system; the trajectory went around one "wing" when the rising heat of the (real) system caused the fluid to roll in one direction, and crossed to the other "wing" when the (real) system reversed itself. In the early 1970s, a Belgian physicist named David Ruelle and a Dutch mathematician named Floris Takens recognized what Lorenz had created and gave it the name "strange attractor."

After other scientists discovered other strange attractors that described other turbulent systems, scientists realized that strange attractors are fractals. If you magnify a small segment of a fractal, you see the same configuration of the entire portion repeated in the magnified segment. Scientists call this characteristic of fractals "self-similarity." The classic example is a coastline—any magnified segment of a coastline shows the same configuration as the entire coastline. The "wings" of the Lorenz attractor look at first glance like almost concentric loops. If you blow up any two adjacent loops, however, you find four loops. If you blow up the four loops, you find eight, and so on.

In a sense, strange attractors allow one to predict the unpredictable—but only to a degree. It's like telling a couple of friends that you'll meet them after work on the running track. They don't know exactly where you'll be at any given moment, but they know that your trajectory will fall within the shape defined by the track.

Can strange attractors and chaos theory be used to better understand complex systems, such as Earth's weather?
(See page 104: Weather Systems)

INFINITY ✓

Infinity goes in two directions: larger and larger, and smaller and smaller. In other words, you can keep on adding more, or you can keep on subdividing. Though infinity is not hard to grasp intuitively, philosophers and mathematicians who think about infinity manage to come up with paradoxes that make the topic more complex than it first appears.

People have long been fascinated by ideas of the infinite. One of the earliest recorded thinkers on the subject was the Greek philosopher Aristotle, who lived from 384 to 322 B.C.E. Aristotle claimed that no group of things could be what he called "actually" infinite, whether infinitely large or infinitely small. Something could only be potentially infinite. In other words, any finite collection of things could potentially have elements added to it. To this day, some philosophers follow Aristotle's belief that infinity is only potential.

Another famous Greek philosopher who pondered the subject was Zeno, who was born about fifty years after Aristotle. Zeno devised a number of paradoxes, known as Zeno's paradoxes. The best known has to do with Achilles and the tortoise. If the slow-moving tortoise is given a head start, how can the swift Achilles ever overtake him—for each time Achilles reaches a point the tortoise has passed, the tortoise will already have advanced beyond it. The paradox is based on the question of how an infinite number of points can be covered in a finite period of time.

The modern-day thinker who changed the face of infinity was the late-nineteenth-century, Russian-born German mathematician Georg Cantor, who discovered what is called the theory of infinite numbers. In defiance of Aristotle, Cantor accepted the existence of the actual infinite. Cantor's way of thinking about infinity goes against our everyday way of thinking about numbers. For example, mathematicians find it handy to pair things. If you take the set—mathematicians call the collection, or group, of something a set—of all integers (whole numbers), it is an infinite set. That is, however high you count, you can always add another integer. If you then take the set of all even integers, it, too, is infinite—you can always add another even integer. Therefore, unless you want to say that one infinity is smaller than another, you are forced to admit that the part (the even integers) is as large as the whole (all integers). Cantor, who accepted this paradox, was comfortable with the idea that there are as many even (or odd) integers as there are all integers. At the same time, he argued that there are different sizes of infinite sets; that is, some infinite sets have more elements than others.

Mathematicians are not the only scientists to think about the infinite. Cosmologists wonder whether or not the universe will expand forever; chemists look at infinitely reversible reactions. But cosmologists and chemists are less troubled by definitions of infinity itself. For mathematicians, Cantor's work left many questions unanswered. Perhaps questions are like integers: you can always add another.

Will time go on forever? (See page 3: Time's Arrow)

Can the decimal form of a fraction go on forever? (See page 67: Irrational Numbers)

THE AGE OF EARTH

Many scientists would respond to the question "How old is Earth?" with the answer, four and a half billion years. But behind the simple answer lie more complex questions. What exactly do we mean by Earth's age (a planet isn't born at a precise moment)? What kind of evidence is available to us? What is the significance of knowing Earth's age?

As it turns out, though Earth wasn't born at a precise moment that could be recorded on a birth certificate, the process of Earth's formation took place over what, geologically speaking, was a relatively short period of time—probably less than 100 million years. Astronomers think that the planets of our solar system formed when dust and ices that were not captured by the newly formed Sun clumped together into small bodies called planetesimals. Some of the planetesimals grew until they became bona fide planets—one of them, of course, being Earth.

One way to determine Earth's age would seem to be to measure the age of the oldest materials, namely rocks. But whatever rocks were around when Earth formed have undergone such drastic changes that they are no longer the same rocks. When convection causes the tectonic plates to collide and pull apart, parts of Earth's crust are submerged into the molten mantle, some of it to be pushed up again as new crust. Other areas are pushed up to form mountain chains. Even rocks that stay where they are worn away by erosion. To try to determine Earth's age by the age of present rocks would be akin to trying to determine a person's age by finding out the age of his or her fingernails.

Because the entire solar system formed together over a relatively brief period of time, scientists can use material from the Moon and meteorites to date Earth's age. (Meteorites are meteoroids that make it through Earth's atmosphere without burning up and land on Earth.) The Moon and meteorites are not altered by the processes of convection, but retain their original material. Thus, they are more accurate indicators of Earth's age than Earth itself.

Scientists measure the age of Moon rocks and meteorites by what they call radioactive dating. In some minerals, the nuclei of the atoms decay at a steady rate, emitting radioactive particles. Scientists refer to the half-life of such a material—that is, the amount of time it takes for half of the material to decay. By comparing the proportion of stable atoms to radioactive atoms, scientists can determine the age of the material. For example, if the half-life of a material were one million years, and three-quarters of the atoms were still radioactive, scientists would know that the material was five hundred thousand years old.

Though it is natural for humans to be curious about Earth's age—after all, most of us are curious about each other's age—by itself the answer is not particularly meaningful. The value of the answer lies in what it can help us to learn about Earth's history—for example, about the separation of the planet into the layers of core, mantle, and crust. But even though the age is not that meaningful by itself, there is something satisfying about just knowing that it's probably about four and a half billion years.

What are the tectonic plates that make it so hard to tell Earth's age from looking at today's rocks? (See page 94: Plate Tectonics)

If our solar system formed over a relatively short period of time, are all the planets similar in composition? (See page 87: Similarities Between Earth and Other Planets)

EARTH INSIDE OUT ——————————————————————— **87**

SIMILARITIES BETWEEN EARTH AND OTHER PLANETS

The planets of our solar system fall into two basic groups: the inner ones, the small solid planets; and the outer ones, the giant gaseous planets. Besides Earth, the small solid planets include Mercury, Venus, and Mars (though not a planet, our moon is sometimes included in this list). The gas giants are Jupiter, Saturn, Uranus, and Neptune. Pluto, the outermost and least-known planet, probably consists mostly of frozen gases.

Venus and Mars are the most similar to Earth. Venus, the closest in size to Earth, has about four-fifths the mass of Earth, and its interior is probably fairly similar to Earth's, with a core, mantle, and crust. Because Mars has only about one-tenth the mass of Earth, there is probably less contrast between Mars's core and mantle. Venus, Mars, and Earth all have atmospheres. The atmospheres of Venus and Mars consist mostly of carbon dioxide, with nitrogen the second most abundant element. The predominant element in Earth's atmosphere is nitrogen, followed by oxygen, which is maintained by the processes of life. Only Earth has oceans. Mars is too cold for large bodies of water to exist without freezing, and Venus is so hot that any substantial bodies of water would evaporate into steam.

Mercury is the outsider of the terrestrial planets. It has no atmosphere, but is covered instead with a layer of fine powder. Mercury is extremely dense, suggesting that its core consists mostly of iron—an even greater proportion of iron than Earth's core.

The gas giants are all very unlike Earth. Jupiter and Saturn, the two largest planets, greatly resemble the Sun in that they consist largely of hydrogen and helium. Though Uranus and Neptune contain large amounts of hydrogen and helium, they also contain other, heavier elements. Both Uranus and Neptune have very deep atmospheres.

Of all the planets, scientists know the least about Pluto. Pluto is a tiny planet, with a mass only about one-fifth that of our moon.

It appears to consist of frozen gases and has an atmosphere that contains methane. Some scientists speculate that Pluto may be a former satellite of Neptune that was pulled into solar orbit by the gravity of a passing body.

Often when people ask about similarities between Earth and other planets it is because they want to know about the possibility of life on other planets. In this case, the meaning of the word *similar* narrows. The conditions on Earth that favor life are very, very precise. For example, if Earth were just slightly farther from the Sun, all the water would be frozen in ice and no life as we know it today could exist. The two planets most similar to Earth are Venus and Mars. But Venus is too hot to sustain life. Mars is the most likely candidate, since it has both an atmosphere and more moderate temperature. However, there is very little moisture on Mars. The most optimistic scenario—for those pleased by the idea of life on other planets—is that previous life forms on Mars were wiped out by what is currently an ice age on Mars. If, in the far distant future, that ice age passes, life could once again flourish on the planet most similar to Earth—Mars.

How do the processes of life maintain Earth's oxygen?
(See page 121: Photosynthesis)

FOSSILS

Throughout the ages, most plants and animals have died, leaving no trace of their (relatively) brief existence behind them. But a very small percentage have been buried under precisely the right conditions, and then left undisturbed long enough for their bodies or body parts to harden into what we call fossils. As fossils, these creatures whose individual lives were so brief are immensely useful to geologists, who study the history of Earth, and to paleontologists, who study the history of life on Earth.

The fossil record, as scientists call the fossils taken together as a whole, is by its very nature partial. Usually, only hard body parts such as bones, teeth, and shells became fossilized. Although soft parts such as feathers and tissues generally decayed soon after death, occasionally they left imprints in rocks. Some of the most extraordinary fossils are those found preserved in ice or oil seeps. In Alaska and Siberia, scientists have found entire mammoths frozen in the ground. In Poland, a woolly rhinoceros from the Ice Age was found preserved in oil seeps, complete with unswallowed food in its mouth!

Many fossils are the result of the organisms having petrified, or turned to rock. This can occur in three ways. In the first way, minerals from underground fill the microscopic voids of the structure. In the second way, minerals replace the original material. Because this happens molecule by molecule, all the details of the original structure are preserved, though not as sharply as in the original. Many fossils are the result of minerals' both filling voids and replacing original material. A third type of rock fossil results when soft tissues consisting mostly of compounds of carbon, hydrogen, and oxygen undergo what is called destructive distillation. Carbon dioxide and water are released until all that remains is a layer of carbon that bears the imprint of the original soft-bodied animal or leaf.

Another category of fossil consists of molds and casts. A mold is the hollow form you would get if you covered your nose with plaster and then removed it; a cast is what you would get if you filled the mold with clay and then carefully removed the mold. Fossil molds and casts formed in similar fashion. A mold formed when solid sediment hardened around an organic object, which then dissolved. If minerals then filled the hole, they formed a cast of the mold. Because footprints are natural molds, they often filled to form casts.

Though individual fossils are fascinating in their own right, scientists mostly study fossils in relation to one another. Geologists, especially, use fossils to date layers of sedimentary rock. Sedimentary rocks, which account for about 75 percent of Earth's solid surface, formed from the breakdown of older rocks from weathering and erosion. The more violent forces, such as volcanoes, that formed other types of rocks would have destroyed any fossils in the process of forming. Because virtually all species of animal become extinct, either through dying off in a mass extinction or by evolving into new species, the presence of a particular fossil in a rock layer tells geologists approximately when that rock was formed. This is especially useful for comparing rock layers from distant geographic areas. If the same fossil appears in two different areas, geologists know that they formed at the same time. Fossils also give clues about geographic and climatic conditions in previous times. If geologists find fish fossils in the rock of a large land mass, they know that at some point in the past there must have been bodies of water large enough to support fish.

Paleontologists, who study the history of life forms, use the geologists' knowledge of the rocks to study evolutionary changes. A primary assumption of geology is that older rocks underlie newer rocks. Therefore, fossils found lower down are older than fossils found above them. By studying the sequence from bottom to top, the paleontologist pieces together the narrative of evolutionary change. This is the same as if a paleontologist of laundry examined a pile of dirty clothes on your closet floor. If a fancy shirt and jacket on the bottom were covered by jeans and a sweatshirt, which in turn were covered by sweaty running clothes, the paleontologist would know that sometime in the past you went to a fancy affair; more recently, you hung around your house or the mall in jeans and a sweatshirt; and most recently, you went running.

Of course, finding fossils is a lot more difficult than finding dirty laundry. Paleontologists must patiently sift through large expanses of rock with no promise of reward. They must handle their delicate findings with extreme care, and then use all their powers of detection and imagination to read the history of life from the remains of creatures long dead.

> *Can paleontologists tell how quickly evolutionary changes occurred?* (See page 107: Evolution: Gradual or Sudden?)

THE CENTER OF EARTH

Except for the occasional earthquake powerful enough to make the front page of the newspaper, we usually think of Earth as solid beneath our feet. Even if we know that plate tectonics causes constant shifting of the continents, we think of it as involving only Earth's outer surface, or crust. But there is movement throughout our planet, from the outer crust to the molten "seas" of the core.

Earth consists of concentric shells, or layers. The outermost layer, the crust, is only ten to twenty-five miles thick. It is the layer of our familiar geography—of our mountains, valleys, and oceans. The most dramatic motions we experience on the crust are the sudden and sometimes deadly earthquakes. Though major quakes seem to strike out of the blue, seismologists constantly work to refine their tracking of the smaller jolts that indicate the mounting strains that eventually cause the "big ones."

Underlying the crust is the mantle, which consists of heavy rocks. Geologists sometimes find surface outcrops of mantle rock. The mantle is about 1,800 miles deep. Beneath the mantle is the core, which has a radius of a little under 3,000 miles. The core has a solid inner kernel surrounded by an outer liquid core of intensely hot molten metals such as iron and nickel. At a temperature of about 12,500 degrees Fahrenheit, the liquid is hot enough to instantly vaporize any known substance. Scientists think that some of this heat was generated by the tremendous compressive forces when Earth formed, and that some of it is from the decay of radioactive materials.

The motions and collisions of the plates that form and support our continents are the outermost manifestation of the heat and motion deep in Earth's center. Just a few miles below Earth's crust, the upper mantle moves slowly in response to the heat of the liquid outer core. The motion in the outer core is not confined to the jostling of the intensely heated molecules. The outer core itself rotates. In some complicated way scientists do not completely understand, this rotation causes Earth's magnetic fields. And if that's not enough motion for you, more than three hundred times in Earth's geologic history, the magnetic poles have reversed. Each time, over a period of a few hundred to a thousand years, the north and south magnetic poles simply swapped places. Even when they're not swapping places, they wander by a few miles each year.

How do we know what is deep inside Earth if we can't go and look at it or take samples? The answer is, through Earth's motions. Seismologists, the scientists who study earthquakes, use their data to theorize about Earth's composition and history. Seismologists use instruments called seismographs to record the sound waves caused by movements such as earthquakes, volcanoes, and explosions from nuclear tests. Because sound waves travel differently through different materials, seismologists use their data to infer what materials the waves have traveled through. Scientists also assume Earth to be similar in structure and composition to meteorites, which they have been able to study.

We still have a lot to learn about Earth's interior. Though Earth's core is much closer to us in distance than any celestial bodies, its intense heat actually makes it less accessible than the Moon.

Do the hot materials deep within Earth ever escape to the surface? (See page 96: Volcanoes)

What are Earth's magnetic fields? (See page 100: Earth as Magnet)

PLATE TECTONICS

When we look at a map, it is easy to see how the eastern border of South America and the western border of Africa appear to fit together like pieces of a puzzle. In 1912, a German meteorologist named Alfred Wegener developed the idea, called continental drift, that said that the reason the pieces appear to fit together is that at one time they were together—they were a single land mass that split apart. Similarities between rocks and fossils on the two continents supported his idea. Though today the idea seems simple and clear, for about fifty years very few scientists took the idea of continental drift seriously, because they lacked an explanation of the forces that would move the continents. Then, in the 1960s, the new theory of plate tectonics elegantly explained continental drift, as well as other phenomena.

According to plate tectonics, Earth's outer surface, called the crust, consists of about ten rigid slabs, or plates. The plates are not the same thing as the continents; they carry the continents, as well as the ocean floors. The plates, which are in constant motion, are the cause of much geologic activity, including mountain formation, volcanoes, and earthquakes. When we say *plates*, we're not talking about dinnerware; the plates that make up Earth's crust are tens of miles thick. We're also not talking about motion that you can feel if you stand still and concentrate really hard. Each plate moves just a few inches a year.

The movement of the plates is the outer manifestation of the roiling of the molten mantle, the layer of Earth that extends from below the crust to the core in the center. Most geologists think that the motion is driven by convection currents, circular currents caused when heated material, which is lighter and less dense, rises and is replaced by cooler material from above. The movement of the convection currents in the mantle carries the plates with it.

The plates undergo three basic types of movements: they move apart from one another; they converge, with one plate becoming submerged below the other; and they slide past each other. In all three types of movement, only the edges of the plates deform; their interiors remain relatively unaffected. When plates separate, or diverge, hot material from the mantle rises up through the gap and solidifies, forming new crust. Since Earth is constant in size, if new crust is formed by separating plates, then somewhere else crust must be taken away to balance things out. This happens when plates con-verge. The plate that slips beneath the other is pushed back into the mantle, where it is melted down by the intense heat. Eventually, it may be pushed up through separating plates to become crust again. If converging plates have land masses on them, then the boundary where they collide crumples, forming mountain ranges. The Alps, for example, were pushed up when Italy joined Europe (geographi-cally, that is, not politically). In the third type of movement, two plates slide or grind against each other, creating an earthquake zone.

Recently, some geologists have questioned whether the move-ment of the plates by themselves could have caused the breakup of the supercontinent 180 million years ago that formed today's conti-nents. They suggest that mantle plumes, plumes of molten rock from the boundary of the liquid core and mantle, shot up through the mantle and crust, causing the ancient supercontinent called Gondwanaland to break apart.

Nothing is permanent, not even the continents. So, wherever you live, even if you don't travel abroad, if you wait long enough, abroad may come to you.

What is the source of the heat that causes the convection movements of the Earth's mantle? (See page 92: The Center of Earth)

Is there another type of convection that affects conditions on Earth? (See page 104: Weather Systems)

VOLCANOES

*B*leb, *slurry, cow-dung bomb, spatter.* These are not comic strip talk, but the words geologists use to describe what comes up and out of a volcano. Unless you live in an area with a lot of volcanoes, you probably think of a volcano as a natural disaster that happens every few years and kills a lot of people. You read the horror stories and lists of agencies to send donations to, and then volcanoes seem to disappear until the next disaster. But though each volcano seems to be an isolated explosion, like the inexplicable fury of a normally composed person, smaller eruptions are relatively common—occurring on average fifty to sixty times a year. There are approximately five hundred active volcanoes throughout the world, "active" meaning a volcano that scientists believe could erupt again at some time. Volcanoes are our means of gazing into Earth's interior; they are also the cause of some of Earth's most dramatic landscapes.

Volcanic mountain ranges are concentrated along belts of seismic activity. One such belt is the "Ring of Fire" around the Pacific Ocean, a segment of which goes from Lassen Peak in northern California to Mount Garibaldi in British Columbia. Mount St. Helens is along this belt. Seismic activity, which also causes earthquakes, is the movement of the large plates that make up Earth's crust. Though we usually feel Earth's crust as solid beneath us, it is constantly moving. Sometimes plates separate from each other; new material from the mantle then rises to the surface, forming a volcanic range. When two plates collide, they can rub against each other, creating a long earthquake fault such as the San Andreas fault in California. Sometimes one plate is forced beneath the other; the submerged plate then returns to the mantle, where it melts down and then rises again as magma in a volcanic burst along the edge of the upper plate.

Magma is molten materials from Earth's crust and mantle. Sometimes it cools and solidifies as rocks within the crust. Otherwise it erupts through a vent in the form of lava. Where lava has cooled and accumulated, we have the volcanic cones that we call hills and mountains.

The material spewed out by a volcano consists of lava, gases, dust, and rock fragments. The gases are the direct cause of the eruption. Because magma is lighter than the rocks surrounding it, it rises towards Earth's surface. As it approaches the surface, gases dissolved in it boil out. This expansion expels the lava and solid fragments from the vent. Lava, which typically erupts at temperatures of 2,100 to 2,200 degrees Fahrenheit, ranges from liquid to viscous. Thin, liquid lava travels the farthest, sometimes tens of miles from the site of the eruption. And as if lava, gases, and rock fragments weren't enough, volcanic eruptions also cause landslides, mudflows, and tsunamis—huge ocean waves.

Not surprisingly, volcanology can be a dangerous area of research. Though some researchers are mainly theorists, others spend a lot of time "in the field." The field is often the rim of an active volcano that is reticent about its plans. In 1993, eight volcanologists were killed in two separate incidents when craters they were studying suddenly erupted. In 1991, a French husband-and-wife team famous for their photographs and films of erupting volcanoes were killed at Mount Unzen in Japan, along with an American postdoctoral fellow. But though some may enjoy the thrill of danger, most volcanologists are motivated by their fascination with a still-mysterious phenomenon—and by their desire, through better volcano prediction, to help save lives.

By the way, a bleb is a small particle of material, slurry is a liquidy mud, a cow-dung bomb is a soft blob in a shape its name suggests, and spatter is fragments of frothy lava that splash as they land.

Where does lava come from and why is it so hot? (See page 92: The Center of Earth)

Why are the plates that form Earth's surface moving? (See page 94: Plate Tectonics)

GROUNDWATER

When we look at a world map, we see landmasses and oceans. When we think about what's inside Earth, we think of rocks and then, further below, molten lava and the core. But Earth's crust contains more than dry rock. Below the surface, even in desert areas, water flows through the rock. Geologists call this subsurface water groundwater.

Most of the groundwater comes from rain and melted snow that seep through the soil. When the water reaches rock, it fills up the pores of the rocks, as well as cracks and spaces between the rocks. The water then flows through this network of openings. The word *flow*, however, does not imply speed: groundwater flows anywhere from about 4 feet per day to 4 feet per year.

Geologists divide the layers of rock that contain the groundwater into two zones. In the top zone, called the zone of aeration, the water shares the rock pores and other openings with air. Below the zone of aeration is the zone of saturation, in which the water completely fills the openings. The zone of saturation rarely extends deeper than 2,000 feet. The upper surface of the zone of saturation is the water table. Generally, the water table echoes the topography of the land above it. In flat areas, the water table is flat, and in hilly areas, the water table rises and falls with the hills. Swamps, lakes, and many streams are actually places where the water table is at the surface—that is, there is no zone of aeration above the zone of saturation.

One way we withdraw groundwater is through wells. A well is just a hole that extends down through the zone of aeration into the zone of saturation. Water flows from the rock pores into the hole, filling the well up to the level of the water table. Too much pumping from a well causes the water table to drop, sometimes hundreds of feet. Because groundwater flows so slowly, it can take hundreds of years to bring a water table back up to its former level.

In some areas, groundwater comes to the surface far more dramatically than in mere lakes and streams. The Thousand Springs, for example, near Twin Falls, Idaho, discharge about 37,000 gallons of water per second. Even more spectacular than springs are geysers, which erupt when the groundwater pressure in caverns or porous rock builds so high that any time the temperature rises, the water turns instantly to steam. When the water at the bottom, which is under the greatest pressure, boils, the steam thrusts the water from below high into the air. After the eruption, the cavern or rock fills with water again and the process repeats. The eruptions of certain geysers such as Old Faithful often seem predictable because the chambers fill rapidly. In geysers where the intervals between eruptions are longer, the many variables make it impossible to predict the eruptions.

Many people have seen the effects of groundwater without realizing it. The eerie-looking stalactites and stalagmites in caves, called dripstone, are the buildup of calcium carbonate deposits from groundwater seeping through cracks. The stalactites, which hang from above, form from groundwater seeping down. Stalagmites then form from calcium carbonate that drips down from the stalactites. As they build up, the stalactites and stalagmites frequently join together in what are called columns. Groundwater performs similar artistry when it seeps through wood, replacing the wood with silica it is carrying. The eventual result is what we call petrified wood. The hard, sculptural shapes of dripstone and petrified wood are reminders that water slowly circulates everywhere below Earth's surface.

What is water made of and why is it so important? (See page 47: Water)

Is there likely to be groundwater on other planets? (See page 87: Similarities Between Earth and Other Planets)

EARTH AS MAGNET

Though Earth doesn't look like a giant iron bar, the planet does act like a single magnet. Every magnet, no matter how small, has two poles. Earth's magnetic poles, which we call north and south, do not exactly coincide with the geographic poles. The quivering little needle on a magnetic compass points not to the geographic North Pole, but to the magnetic—a fact of scientific interest, and no small significance if you're wandering in the woods lost on a camping trip.

Earth's outer liquid core contains iron and nickel. In magnetic materials such as these metals, individual atoms align themselves along the magnetic axis. Eventually, the random motion of the atoms causes them to lose their alignment unless some other force keeps them aligned. The outer liquid core is in constant motion relative to the solid inner core. Scientists think that this motion causes currents that generate electricity, which in turn generates a magnetic field. Because this magnetic field extends outward from Earth's surface for hundreds of miles into space, we say that Earth acts like a single giant magnet.

To scientists, Earth's magnetic behavior is interesting not only in itself, but for what it explains about other phenomena. The French physicist Pierre Curie discovered that above a certain temperature, later called the Curie temperature, a metal loses its magnetism. The reverse of this is that, as a molten metal cools, it becomes magnetized when it reaches its Curie temperature. It didn't take paleontologists long to realize that this characteristic of metals could be very useful. Because lava cools fairly quickly—within a matter of weeks—the alignment of metal grains in volcanic rock shows how they were aligned when the lava cooled to its Curie temperature. They are like little magnets frozen in time. One of the first surprises learned from these frozen magnets was that Earth's magnetic poles have reversed many times throughout geologic history. The most recent reversal occurred about 730,000 years ago. Scientists are not sure whether the poles weaken and then reverse, or whether they reverse suddenly—suddenly meaning over a period of 1,000 to 5,000 years. Each magnetic epoch, as it is called, lasts millions of years; within these epochs occur shorter reversals, called magnetic events, that last a mere several thousand to 200,000 years.

These magnets "frozen" in rock also provide evidence for the theory of plate tectonics—the idea that Earth's crust consists of rigid slabs, or plates, whose constant motion is the cause of all geologic activity, including the movement of the continents. Where plates diverge, new molten material rises up from the mantle below and hardens. Eventually, this material, too, separates, to be replaced by even newer material from the mantle. Along the ocean floors, reversals of Earth's magnetic poles have caused striped patterns in the rocks on either side of the mountain chains called mid-ocean ridges. The divergence of these ridges causes what scientists call the ocean floor spreading. The movement of the ocean floors, at the rate of a couple of inches a year, is an important part of the overall movement of the tectonic plates.

For now, you can rely on your pocket compass to get you out of the woods. Of course, after the next reversal, it'll still get you out of the woods—as long as you remember to line up the needle with the S instead of the N.

How old is Earth, that it has had many lengthy magnetic epochs? (See page 85: The Age of Earth)

THE CORIOLIS FORCE

The first thing to know about the Coriolis force is that it is not a force (a force is an influence on a body that causes it to accelerate, such as the force of a person pushing a stalled car). The second thing to know about the Coriolis effect, as it is more accurately called, is that it does not cause bathtubs in the Northern Hemisphere to drain counterclockwise and bathtubs in the Southern Hemisphere to drain clockwise; its effects are visible on much larger systems, such as atmospheric and ocean currents.

The Coriolis effect, which in both hemispheres deflects objects eastward as they travel toward the poles, and westward as they travel toward the equator, was discovered by a French engineer and mathematician named Gaspard-Gustave de Coriolis (1792–1843) as he studied the dynamics of rotational machinery. The concept is fairly straightforward. Though it applies to all rotational systems, it is clearest if we look at how it operates on Earth. Earth completes a rotation once every twenty-four hours. Because the length of Earth's circumference, or equator, is greater than that of a circle closer to either of the poles, an object at the equator travels a greater distance in each twenty-four-hour period than does an object closer to a pole—approximately 24,000 miles at the equator and zero miles at the poles themselves. Thus, an object at the equator travels eastward at a velocity of about 1,000 miles per hour, whereas an object at either pole doesn't travel at all. If an object such as a rocket leaves the equator and heads toward the North Pole, inertia will cause the rocket to continue to travel eastward at about 1,000 miles per hour, ignoring, for simplicity's sake, the velocity of the rocket itself. (Inertia is the tendency of an object at rest to stay at rest, and of an object in motion to stay in motion.) When the rocket, still traveling 1,000 miles per hour, reaches a latitude that is traveling, say, 500 miles per hour, the rocket's greater velocity causes it to be deflected eastward.

Sometimes people illustrate the Coriolis effect with examples of phonograph turntables and merry-go-rounds. A line drawn from the center of a turning record to its rim will be deflected at an angle, as will a person who walks from the center of a merry-go-round to its edge. However, it is important not to confuse the Coriolis effect with the simple displacement caused by rotation; in other words, as you approach the leprechaun at the edge of the merry-go-round, the leprechaun itself moves. In fact, it's doubtful that the Coriolis effect has any more effect on a turntable or merry-go-round than it does on a bathtub drain.

Most of us don't have to worry about aiming rockets, or even piloting planes, but we do worry about weather patterns—even if we can't do anything about them. A clear example of the Coriolis effect is the fact that cyclones, hurricanes, and tornadoes tend to spin counterclockwise in the Northern Hemisphere and clockwise in the Southern Hemisphere. The effect is negligible on smaller weather systems such as thunderstorms.

The Coriolis effect is present on all rotating bodies. Perhaps the most dramatic example is Jupiter's Great Red Spot, which is nothing more (or less) than a huge storm spinning counterclockwise in Jupiter's southern hemisphere.

How does inertia affect a moving object, such as a rocket? (See page 29: Inertia)

What causes weather? (See page 104: Weather Systems)

WEATHER SYSTEMS

It's not the heat; it's the humidity." Though the cliché is true, our weather is a combination of many factors, including heat, humidity, clouds, winds, pressure, precipitation, and visibility. These factors all describe the condition of the atmosphere, the layer of gas that surrounds Earth's surface—planets and moons that have no atmosphere have no weather.

When we turn on the radio in the morning, most of us just want to know how hot or cold it's going to be, and whether it's going to rain or snow. But to the meteorologist, these elements are part of a larger structural picture, called weather systems. Weather systems are the result of the constant changes of the atmosphere, which in turn are the result of constant changes in temperature.

A phenomenon called convection is responsible for all the air movements that determine our basic weather patterns. When air is heated from the warmth of Earth's surface, it expands and becomes less dense. Because it is less dense, and thus lighter, it rises; colder air rushes in below to replace it. This air in turn grows warmer and rises, again to be replaced by colder air. The constant exchange causes circular movements of air. Because the air at the equator is warmed more rapidly than air at the cold poles, it rises more rapidly and is replaced by colder air that rushes in from the north and south. Earth's rotation complicates the pattern. The distance around the equator is much greater than the distance around Earth near the poles. Because Earth completes one rotation every twenty-four hours, air at the equator must cover a greater distance, and thus is traveling much faster than air at the poles. This difference in speed causes the air currents to be deflected. Scientists call this deflection the Coriolis effect. Air traveling from the equator toward the poles shifts eastward, and air traveling from the poles toward the equator shifts westward.

The major flow of air over the United States is called the prevailing westerlies. The prevailing westerlies blow west to east in the temperate zones of the Northern and Southern Hemispheres. In the Southern Hemisphere, instead of flowing north toward the North Pole, the winds flow south toward the South Pole. In regions closer to the equator, the flow is reversed, with the westward-flowing northeast trades (above the equator) and the southeast trades (below the equator). At the equator itself, there is often a mass of stagnant air called the equatorial doldrums.

Earth's surface is, of course, not uniform. Land masses reradiate absorbed solar energy more rapidly than do bodies of water. These temperature differences cause large whirling masses of air called high- and low-pressure cells, which generally alternate. High-pressure cells usually bring calm, pleasant weather, and low-pressure cells bring storms.

Though the general weather patterns are fairly consistent, our local weather, the weather we turn on the radio to hear about, is also influenced by many other factors. Among the most important are air masses, large moving bodies of air that can cover hundreds, even thousands of square miles. As an air mass travels, it takes on the temperature and humidity conditions of the surface over which it travels. Meteorologists call the collision of two air masses a front. Fronts usually cause bad weather. Weather forecasting basically consists of predicting the activity of air masses.

Other factors that influence the weather are seasonal changes; the jet streams, which are narrow, high-altitude wind currents that speed from west to east in the Northern Hemisphere; and local changes in temperature, pressure, and humidity. Other than temperature, the aspect of weather we're most interested in is the water cycle, a fancy term for whether it's going to rain or snow. Every day, heat evaporates water from Earth's surface into the air. As the heated air rises, it cools, and vapor condenses into clouds. Under certain atmospheric conditions, the clouds produce rain or snow.

The dynamics of weather are so complex that scientists consider weather to be a classic example of chaos theory, the study of unpredictable, turbulent systems. Because of weather's extreme complexity, it is impossible to make forecasts with any accuracy more than several days in advance. Weather may be inconvenient, but at least it gives us something to talk about.

What is chaos theory, and why is it useful in the study of weather systems? (See page 79: Chaos)

Convection causes movements in the air, but can it also cause movement in Earth itself? (See page 94: Plate Tectonics)

EVOLUTION: GRADUAL OR SUDDEN?

Virtually all scientists agree that Darwin's theory of evolution is as reliable a description of the world we live in as is Newton's law of gravitation or Mendeleev's periodic table of the elements. Any object you drop will fall to Earth, and any new chemical element that is discovered will fit into the periodic table. Just as surely, any animal you come across has ancestors in the evolutionary past. But as in any scientific field—and it's part of what keeps science exciting—there are areas of evolution about which scientists disagree sharply. The main argument, which has been going on for almost twenty-five years, is over the rate of evolution; that is, whether evolutionary changes occur gradually or in sudden intense spurts. The former viewpoint is called gradualism, and the latter is called punctuated equilibrium.

Darwin himself thought that evolution proceeded at a slow, even rate, and that small changes gradually accumulated until they had been transformed into big changes. Some think that one reason he saw evolutionary change as gradual is that he was strongly influenced by the field of geology. It may have seemed natural that if Earth's geology changes slowly, so would the organisms that inhabit it.

Armed with contemporary knowledge about DNA, the material that carries our genetic code, some of today's gradualists go a step beyond Darwin. They say that although the physical characteristics of a species may evolve at a somewhat irregular rate, changes in the DNA appear at a uniform rate. They call this the molecular clock and claim that by counting the number of changes in species' DNA, you can determine how much time has passed since different species diverged from common ancestors.

Punctualists, as those who believe in punctuated equilibrium are called, say that species remain basically unchanged for as long as one million years and then undergo brief periods of intense change. It is during these periods of change that new species diverge from old ones. Punctualists often point to the lack of fossils showing intermediate stages of change as evidence in their favor. Paleontologists—the scientists who study fossils—used to assume that they just hadn't yet found the fossils displaying the intermediate stages; punctualists say that that may be because they never existed.

A major problem in resolving arguments about the rate of evolution is finding the evidence. Although scientists can determine that one animal evolved from another, it isn't possible to just go out and find an animal in the act of evolving, the way one can find an animal in the act of eating or catching prey. The main source of evidence for evolution is the fossil record. Because older fossils are embedded under more layers of rock than are newer fossils, paleontologists can often follow the evolutionary history of an organism by studying how its fossil remains changed over time. But the fossil record is sketchy at best. Usually, only hard body parts such as bones fossilize; soft parts decay. And, of course, animals don't always cooperate and form fossils in neat piles one atop another. The same fossils that show sudden change in one rock area may show more gradual change elsewhere.

The rate of evolution may turn out to be both gradual and punctuated. Perhaps evolution occurs more rapidly during periods of great environmental pressure, as when new species of insects evolve in response to pesticides. In the meantime, the paleontologists continue to dig and argue.

Are extinctions also recorded in the fossil record? (See page 111: Disappearing Frogs)

BIODIVERSITY

In our daily lives, most of us are consciously aware of very few species other than humans: domestic cats and dogs, perhaps mice and rats; house flies, silverfish, and earthworms; broccoli and cauliflower. But wait, "earthworm" isn't a single species; there are about 12,000 known species of earthworms. Basically, a species is a category of plant or animal that can breed within itself and produce fertile offspring. Though a horse and donkey can breed, for example, because they are separate species, their offspring, mules, are sterile. Earth contains many millions of species, many of them yet to be discovered. By examining just a single handful of dirt, scientists can discover thousands of new species of bacteria and viruses. To a great extent, the larger the animal, the fewer the species. There are only about 4,000 species of mammals, compared to, say, the 750,000 known species of insects—and entomologists think there may be as many as 10 million species of insects.

Although the specifics of their definitions vary, ecologists generally call the incredible array of species *biodiversity*. Although they are still exploring the details, they already know that biodiversity is crucial to Earth's healthy functioning. Researchers have found, for example, that the more species of plants an area has, the more efficient the area is in making use of the Sun's energy. Often diverse areas are also better able to resist and recover from stresses such as droughts.

The history of life on Earth is marked by what scientists call mass extinctions. These are periods, up to several million years long, characterized by such extreme environmental change or disaster that many species don't have time to adapt and consequently die off. The most famous mass extinction is the one that wiped out the dinosaurs. Some scientists think that we are currently in the process of a new mass extinction—one caused by human activities, including industry and agriculture. Some scientists even think that this may be one of the largest mass extinctions ever to occur on Earth.

Many scientists and conservationists concerned about biodiversity focus their efforts on the tropical rain forests, for two reasons. First, the rain forests are extraordinarily diverse, containing as many as half the world's species. Second, through logging and clearing, the rain forests are disappearing at an alarming rate—an area as large as the state of Florida each year. Although many people support the idea of protecting biodiversity for both philosophic reasons (we shouldn't dilute the richness of the natural world) and practical ones (the human race needs it to survive), it can be hard to convince those who benefit financially from destroying the rain forests to change their ways. One tactic is to search for new ways to profit financially from the rain forests without harming them. In Costa Rica, for example, a National Biodiversity Institute is searching for as-yet-undiscovered species that could be used to produce pharmaceuticals. Though the chances of making such a discovery are very slim, the payoff could be in the millions of dollars. At least one large pharmaceutical company is interested enough to fund the institute's research in exchange for first looks at any potentially valuable compounds that are found.

Even for those whose only environmental concern is human survival, the question is, how many species can we afford to lose before the ecosystem that supports human life collapses? Although there is some redundancy of species, this is one instance in science where we don't want to come too close to finding out the answer from direct observation.

Could the human species unknowingly orchestrate its own extinction? (See page 117: Human Population)

DISAPPEARING FROGS

Although some people find frogs cute, many find them unappealing or even a nuisance. Thus, it's no surprise that most people are unaware that the frog may be on the way to extinction. Everybody loves a panda, and there's a sense of romance about wolves and bald eagles, but who over the age of ten cares about frogs? Biologists who study amphibians (amphibians are a class of vertebrates that includes frogs, toads, newts, and salamanders) do—and in recent years many of them have become alarmed that frogs are on the way to extinction.

Because they are hairless and have permeable skin, frogs are particularly vulnerable to environmental changes. Some people point to the frog as a bioindicator, like a canary in a mine. Miners used to take canaries down into mines to test for gas leaks—if the canaries passed out, dangerous gas was in the underground air. Strictly speaking, however, there must be a direct correlation between a bioindicator and a specific danger. The disappearance of frogs may indicate general ecological trouble, but it does not point to one specific danger.

Frogs are definitely disappearing. For example, the gastric brooding frog of Australia apparently became extinct in 1980. The golden toad of the Monteverde Cloud Forest in Costa Rica became extinct, or nearly extinct, in the 1980s; populations of the mountain yellow-legged frog and the Yosemite toads, both from California, have also declined noticeably. All animals are dependent on the ecosystem they inhabit. Scientists think that the main reason for the decline in frog species is human alteration of the environment. In Great Britain, for example, a combination of pesticides, acid rain, and lack of undergrowth caused the frog population to decline dramatically.

Commercial factors play a role in the frog's fate. India used to be the largest supplier of frogs' legs to Europe (the French, who are fond of eating frogs' legs, passed laws protecting frogs). In some areas of India where frogs had disappeared, pests such as mosquitoes and tiny rice-eating crabs, normally kept under control by frogs, flourished. When India finally banned the export of frogs, Indonesia took its place as Europe's main supplier—and now Indonesia's frogs are disappearing.

As frightening as the word *extinction* may sound, extinctions are "natural." Most species throughout evolutionary history have eventually become extinct, to be replaced by others. We know this from the fossil record. Not only do individual species become extinct, but periodically Earth undergoes what scientists call mass extinctions. During the largest mass extinction, 245 million years ago, as many as 90 percent of the species on Earth became extinct. During the mass extinction 65 million years ago that wiped out the dinosaurs, about 65 percent of the species became extinct. Because mass extinctions take place over a period of up to several million years (brief by geologic standards), it would never occur to most people that we may be in the midst of another mass extinction.

If extinctions are "natural," why should we worry? For one, out of self-interest. Most of us are fond of our species and would like to see it continue. Though we could undoubtedly survive the extinction of many species, the sign of a healthy and vital environment is its level of biodiversity. Humans, who are largely responsible for the rapidly diminishing biodiversity, must now make it a priority to try to undo the damage.

What is biodiversity and why is it important? (See page 109: Biodiversity)

Why did the dinosaurs die out? (See page 115: Why There Couldn't Be a Lone Dinosaur or Two Alive Somewhere)

THE FUTURE OF HUMAN EVOLUTION

When we think of the past, many of us think, perhaps, of Martha and George Washington, or of Michelangelo or Joan of Arc. We think of people who looked basically like ourselves but wore funny clothes and hairdos. When we imagine the future we tend also to picture people who look like us—perhaps flying around in little spaceships. But could our relatives of the future differ from us as much as we differ from our Neanderthal ancestors of 30,000 to 120,000 years ago? Could humans turn into ET-like creatures—large-headed, hairless, and weak-bodied?

The general view of human evolution is from the hulking, hairy, none-too-bright Neanderthal to the trimmer, less hairy, smarter human of today. Though most of us don't identify very strongly with the Neanderthals—we usually use the term *Neanderthal* as an insult—in one respect they were surprisingly similar to us: brain size. In fact, there's been no increase in the size of the human brain for the past 100,000 years. The Neanderthals even had a greater cranial capacity than modern humans. As for hairiness, there may indeed be a trend toward less hair. But the hair of our remote ancestors is not preserved in the fossil record, so we have no way to confirm such a trend. Though we may be less strong than the Neanderthal, we are taller and heavier, as modern populations have enjoyed better health and nutrition. While that trend will probably continue as more peoples become westernized, it's not possible to project reliably into the long-term future.

The theory of evolution says that, like all living creatures, we must be evolving. According to the theory of evolution, random mutations appear, and those best able to adapt to the environment reproduce the most successfully. But the fact that we don't know how the environment will change makes it very difficult to predict how humans will change. For us to change into something as different from ourselves as ET-like creatures would probably entail the development of a new species. Sometimes a new species develops when a species gradually accumulates so many genetic changes that it eventually becomes a new species. Another way is when two or more groups of a single species are geographically—and thus genetically—isolated from each other. They develop different characteristics in response to different environmental pressures over many generations until finally they become different species. One important factor, however, makes any dramatic change through the latter scenario unlikely: our geographic mobility. In today's world of modern transportation and communication, no place is completely isolated. Even the remotest areas are visited by explorers and tourists. Native inhabitants of such areas live elsewhere for a while, and then return, bringing back new ideas, new medicines, and new junk foods. As different as we are from one another in physical and cultural detail, biologically we are one big family, and that is less likely to change as time goes on.

One change in the future may provide the isolation necessary for a new species to develop: space travel. If humans colonize other planets, in this solar system or even in other solar systems, a group of human space emigrants could eventually become an outpost of ET-like creatures. More likely, however, they wouldn't look anything like ET. For the course of evolution is unpredictable—and why should the outcome fit the images of our movies?

Could the human species be wiped out by a disease that had become resistant to our medicines? (See page 125: The Growing Resistance of Bacteria to Our Medicines)

Could the human species overpopulate to extinction? (See page 117: Human Population)

WHY THERE COULDN'T BE A LONE DINOSAUR OR TWO ALIVE SOMEWHERE

Sixty-five million years ago, at the end of the Cretaceous period, the dinosaurs, along with many other species, died off in a mass extinction. Scientists aren't sure what caused the extinction, but they are building a pretty good picture. There are two basic kinds of extinctions: those caused by external factors and those that result when a species changes so much that it evolves into one or more species. Most scientists agree that external factors wiped out the dinosaurs, and that at least one of these was an asteroid that struck Earth. The asteroid itself didn't knock the dinosaurs and other animals dead, but ash and dust from the impact blocked out the sunlight for a number of years. In 1990, scientists dated a large crater in Mexico to the precise time of the dinosaurs' extinction. (Of course, the word *precise* has a different meaning when we refer to something that occurred sixty-five million years ago than it does when we speak of something that occurred two weeks ago.) The crater, which is 180 kilometers in diameter, is the largest on Earth. If an asteroid was responsible for wiping out the dinosaurs, that would help to explain why some animals and plants were able to survive the mass extinction. Smaller animals would have burrows in which to escape the cold, and many plants can survive long periods of time in dormant states.

Some scientists think that even if an asteroid did strike Earth, it was only the final blow to the already weakened dinosaurs. One theory is that lowered sea levels caused previously separated land masses to be joined. When animals migrate from one environment to another, they bring with them new diseases and parasites, which can be deadly to the native population. Other scientists have proposed that as mammals evolved, they ate the dinosaurs' eggs, hastening their extinction. However, mammals existed during most of the hundred million years that the dinosaurs dominated Earth.

The reason no dinosaurs could have survived to today is related to how they, and all other species, developed in the first place: through natural selection. Individuals within a species are born with mutant, or novel characteristics. Most mutations die out. But an individual with a mutation that makes it better adapted to the environment has a greater chance of survival. It produces more offspring, which in turn produce even more offspring. In periods of catastrophic environmental change, the normal rate of mutation—hence, of natural selection—is too slow to allow a species to adapt to the changes. Had any dinosaurs been genetically flexible enough to adapt to the changing environment, they would have had a great survival advantage over the species that could not adapt and became extinct. Thus, they would have flourished and would still dominate life on Earth. Scientists also rule out the existence of a lone surviving dinosaur because a species cannot survive without enough members to provide genetic diversity (too much inbreeding is bad) and a sufficient supply of mates. But as much as we may enjoy our fantasies of finding a lone dinosaur off in a swamp somewhere, we're lucky that they did become extinct. Otherwise, humans may never have made it as a species, regardless of our big brains that can reconstruct dinosaur skeletons and come up with scientific theories.

If an asteroid wiped out the dinosaurs, from where did it come? (See page 9: The Asteroid Belt)

HUMAN POPULATION

As long ago as 1798, the English political economist Thomas Malthus sounded the alarm in his *Essay on the Principle of Population*, claiming that population tends to increase more rapidly than the means of subsistence. He warned that unless population growth was checked either by social planning or by disasters such as famine, disease, or war, our planet would suffer widespread poverty and destitution. Many historians like to say that Malthus's essay sparked Darwin's concept of natural selection—that certain genetic variations tend to survive and others to die off. (Though Darwin's idea is based on random mutations, in popular culture, it's become the one-size-fits-all salute to competitiveness, "survival of the fittest.") One prominent historian, however, says that Darwin was so close to formulating his concept that almost anything he read at the time would have had the same effect.

Scientists speak of Earth's "carrying capacity"—how many human beings Earth can support. Though the concept is relatively simple, two factors make thinking about it more complicated. First, estimates of Earth's carrying capacity vary widely. Recent estimates range from around 3 billion (which we've already surpassed) to 44 billion. Some scientists narrow it to 7.7 to 12 billion. This range is close to UN projections of the world population in 2050: 7.8 to 12.5 billion. The second thing that complicates the picture is that Earth's capacity itself changes, with changes in such factors as food production.

To understand the relationship between population levels and resources, social scientists examine such variables as life expectancy, literacy, income levels, and technology; some compare more specific factors such as carbon dioxide emissions. No matter how you describe the situation, though, most researchers think that if our planet's population continues to expand, and its resources to be depleted, we're bound to end up in trouble.

How fast is the population growing? From 1 C.E. to around 1970, the annual rate of increase went from 0.04 percent to 2.1

percent. From that peak, it has declined to about 1.6 percent. Since 1.6 percent of anything doesn't sound like such a big deal, it's more impressive to look at the numbers themselves. The world's current population is about 5.7 billion people. If it continues to grow at an annual rate of 1.6 percent, it will double over the next forty-three years—to more than 11 billion. The gap between the rich and the poor has grown even more rapidly.

Many things, looked at closely, become topsy-turvy. One might think that countries whose populations are growing quickly pose the greatest threat to the environment. However, the countries with the lowest rates of population growth tend to be the wealthiest—and wealthy countries consume energy at a much higher rate. Poorer people, however, are often forced to farm substandard lands and to deforest hills, which may damage the long-term health of the environment.

Though technology itself is not inherently good or evil, it arouses passions on the part of both its proponents and detractors. Until recently, our lack of awareness of the potential for human impact on the environment allowed us to recklessly assault our surroundings with pollutants and to squander resources with no thought for tomorrow. But by controlling technology, countries can undo some of the damage that's been done, minimize future damage, and lower energy consumption. This requires more than convincing the person who likes to drive a gas-guzzling car to change his or her ways.

What is the greenhouse effect, and how does human population affect it? (See page 123: The Greenhouse Effect)

How do human population growth and resource use affect biodiversity? (See page 109: Biodiversity)

CARNIVOROUS PLANTS

The idea of a meat-eating plant gives a person—at least a non-botanist—a chill. It suggests consciousness, as if the plant intentionally set out to trap and devour prey. Of course, intention has nothing to do with it; it's all a matter of plant structure and chemistry. The carnivorous plants have adapted to the acidic, waterlogged soil of the bogs they inhabit. To meet their nitrogen requirements, they have developed ways to absorb nitrogen from insects and other small animals they catch. The vast majority of carnivorous plants in North American grow in a swath of land that stretches from Virginia to Texas, particularly along the Gulf Coast. Three of the best known are the Venus flytrap, the sundew, and the pitcher plant.

The Venus flytrap, which Darwin thought one of the most wonderful plants in the world, has nectar-producing glands on its leaves to attract unsuspecting insects. Despite its name, its repasts consist of spiders and ants more often than flies. Each leaf has two halves and a hingelike movement that enables them to close. On the leaf are hairs that serve as triggers: if a victim touches a single hair twice or two successive hairs within a period of twenty seconds, an electrical signal shuts the trap door. The closed leaf then secretes enzymes that digest the victim.

The sundew is a small herbaceous plant with tentacles that secrete a sticky fluid that traps any insect foolish enough to come calling. The tentacles then bend around the insect, pressing it against the surface of the leaf to be digested. The plant derives its names from the droplets of fluid secreted by the tentacles, which are said to glitter like dew in the morning sun.

The dramatic-looking pitcher plants have long, hooded tubular leaves that somewhat resemble pitchers. Nectar on the hood and rim of the pitcher attracts insects—and occasionally centipedes and scorpions. Once an insect alights on the slippery rim, it usually plunges to its doom at the bottom of the pitcher. To ensure that no one who enters escapes, the inside of the tube has hairs that point downward, blocking any upward travel. The bottom of the pitcher is filled with water that contains digestive juices and, in some species, a "drug" to stun the victim.

As inhospitable as the pitcher plant may be, a few creatures are able to form mutually beneficial relationships with it. Certain species of insects—as many as sixteen in North America—spend their larval stage within the pitcher, feeding off the bodies of insects in the digestive pool. Scientists aren't sure how these insects survive the pitcher's digestive juices, but think that their bodies may produce protective anti-enzymes. As the larvae feed on the insect carcasses, they help break them down into the nutrients the pitcher plant requires. They also produce their own nitrogen-rich wastes.

Unfortunately, the bogs that the carnivorous plants call home have not been faring well. Some of them have been drained and converted to pine plantations or to farms. Others have been the victim of good intentions. Unless they are pruned by burning, local shrubs and trees invade the bogs and absorb the available water. In the 1930s, the United States Forest Service—and its spokesbear, Smoky—began its campaign against fires, in both forest and nonforest areas. The absence of frequent fires has caused the bogs to decline. A more direct encroachment on the health of the carnivorous plants is their harvesting by florists, who find a ready market for their exotic beauty. After all, to their human admirers, these carnivores pose no threat.

How long would it take a plant to evolve into a meat-eater? (See page 107: Evolution: Gradual or Sudden?)

Does a meat-eating plant still need sunlight? (See page 121: Photosynthesis)

PHOTOSYNTHESIS

Many scientists conduct research solely for the satisfaction of understanding how the world works. Though they know that down the line their results may find practical application, that is not their primary motivation. But occasionally an area of research is so closely tied to an application that it is hard to separate the two. One such field is the study of photosynthesis. At the same time scientists are working to understand photosynthesis, they are searching for ways to create molecules that can do what plants do—convert the Sun's energy into chemical energy while removing carbon dioxide from the atmosphere. A synthetic source of photosynthesis would provide energy by using cheap and abundant resources (light, water, and carbon dioxide). It would also help lower carbon dioxide levels, which contribute to the greenhouse effect.

Photosynthesis, which is essential to all life on Earth, consists of two main processes. In the first, plants take the energy of light and convert it into chemical energy. In the second, called dark photosynthesis (because it can take place in the dark), the plant uses this chemical energy to convert carbon dioxide and water into carbohydrates and oxygen. The carbohydrates provide the food that all animals, including humans, depend on, and the oxygen provides virtually the entire oxygen supply that all animals, including humans, require for respiration. As oxygen producers, plants are no slouches. A single cell of green algae, for example, can produce up to thirty times its own volume in oxygen per hour.

Photosynthesis begins with small structures, called chloroplasts, that contain chlorophyll, the pigment that makes plants green. Chlorophyll molecules capture the Sun's light in ways that scientists don't fully understand. When each light particle, called a photon, hits a molecule of chlorophyll, it "excites" an electron, which sets off a cascade of electrons that leap from molecule to molecule. The final product is the creation of a molecule—with a very long name, shortened to its initials, NADPH—that stores the electrons. Along the way, the electrons lose some of their energy, which is collected into a compound with almost as long a name and the initials ATP.

In the next stage, dark photosynthesis, energy from the NADPH and ATP molecules converts carbon dioxide and water into the carbohydrates the plant needs to grow. Because the chlorophyll needs to replace the electron it lost when it was hit by the first photon, catalysts in the plant cause water molecules to give up electrons to the chlorophyll. In so doing, the water molecules split apart into hydrogen and oxygen, and the oxygen is released into the atmosphere.

The main strategy of scientists trying to produce synthetic photosynthesis is to come up with nonbiological molecules that can mimic the process of setting off the cascade of electrons. In plants, an as-yet-unknown mechanism keeps the electrons moving in one direction. Any system devised by scientists has to be able to prevent the electrons from returning to the molecule they came from. Though a marketable artificial photosynthesis system is probably a number of years down the road, eventually the term *artificial plants* may refer to something other than silk flowers and plastic ivy.

How do increased levels of carbon dioxide in the atmosphere contribute to the greenhouse effect? (See page 123: The Greenhouse Effect)

Is there a way we can convert the Sun's energy into electrical energy? (See page 31: The Photovoltaic Cell)

THE GREENHOUSE EFFECT

Some names that scientists assign to phenomena are merely fanciful—for example, the terms *color*, *taste*, and *charm* as applied to subatomic particles. But other names describe the phenomenon perfectly. The term *greenhouse effect* is an example of the latter. Certain gases in the atmosphere, particularly carbon dioxide, act like the panes of glass in a greenhouse; they allow the Sun's light to penetrate Earth's surface, but then trap the heat that is radiated back from Earth, preventing it from escaping into space. The burning of fossil fuels—coal, oil, and natural gas—is the main cause of the increase of carbon dioxide in the atmosphere. The fuels are called fossil fuels because they form from the decayed remains of organisms. As they burn, the carbon in them bonds with oxygen to form carbon dioxide.

Although the concept of the greenhouse effect is easy to understand, its reality and potential consequences are hotly debated—within both the scientific and the political communities. It's much harder to determine cause and effect in the real world, which is filled with so many variables that cannot be controlled, than it is in the laboratory. Very few, if any, scientists dispute that the concentration of carbon dioxide in the atmosphere has risen 30 percent since the 1880s, and that the temperature of Earth's atmosphere has risen from 0.5 to 1.25 degrees Fahrenheit in the last century. What they do dispute is whether or not there is a cause-and-effect relationship between the two. Some scientists have suggested that the rise in temperature may be due to normal long-term fluctuations in the Sun's brightness. Others claim that the initial result of any greenhouse effect would be increased cloudiness, which would shield Earth from the Sun's light and prevent warming.

Scientists base their projections of the consequences of the greenhouse effect on computer models of the climate. Besides an overall rise in temperature, these models suggest that a greater percentage of rain and snow would fall during short, intense storms, but that there would be less variation in the day-to-day weather.

Scientists have evidence that both phenomena are already occurring. From 1984 through 1994, in the United States alone, there were higher rates of drought, snow, and heavy rainstorms.

If there is a greenhouse effect and it goes unchecked, the consequences could be dire. Although scientists think that total melting of the polar ice caps is unlikely, even a partial melting could cause sea levels to rise, inundating coastal areas and destroying towns. There would be an increase in hurricanes and other tropical storms.

Different computer models give different projections, and those who doubt the existence of the greenhouse effect point to the contradictions of the models. Recently, however, there has been a shift toward taking the greenhouse effect seriously, even by those who were sitting on the fence until now. Rather than asking if there is a greenhouse effect, more scientists—as well as those in industry and government—are asking what should be done about it.

Can plants help combat the greenhouse effect? (See page 121: Photosynthesis)

THE GROWING RESISTANCE OF BACTERIA TO OUR MEDICINES

More than 2,500 years ago, the Chinese used moldy soybean curd to treat infections. You could call this the first clinical use of antibiotics. In our century, Alexander Fleming discovered a crude form of penicillin in 1928, but it wasn't until the 1940s that antibiotics came into widespread use. Optimists thought that the advent of antibiotics heralded a new era in public health and the defeat of such infectious diseases as tuberculosis and pneumonia. They were almost right. For a few decades, many diseases that previously struck fear in people's hearts were controlled with standard courses of antibiotics. But then the bacteria that cause the infections began to outsmart the scientists; they became resistant to our antibiotics.

Infectious diseases are caused by bacteria, viruses, and sometimes fungi. Many people confuse bacteria with viruses—parasitic organisms that cause diseases ranging from the common cold to AIDS. Basically, we have no treatment for viral infections. Bacteria are single-celled organisms that, unlike viruses, can reproduce independently of a host. They cause many common infections, such as bronchitis, sinusitis, and meningitis. Bacteria (and fungi) develop chemicals to kill off other organisms. Humans cultivated these "killer microbes" and used them against specific bacteria. Penicillin, for example, comes from the fungus *Penicillium*. However, scientists have created most of the newer antibiotics in the laboratory.

Antibiotics use a variety of means to destroy or inhibit the growth of bacteria. They prevent cell walls from forming in new bacteria; they erode the cell walls of existing bacteria, causing the bacteria to dissolve, like the Wicked Witch of the West in the Wizard of Oz; they prevent proteins from forming within the bacterial cell; they block the DNA from reproducing new bacteria; and they block different stages of the metabolism of the bacteria.

The fact that antibiotics use these different ways to disable or destroy bacteria means that the bacteria, in turn, have just as many ways to outwit the antibiotics. By outwit, we don't mean that an individual bacterium thinks hard and comes up with a new trick. Instead, in line with evolutionary theory, mutant bacteria with the ability to resist antibiotics survive and reproduce more successfully than those that lack that ability. Penicillin, for example, destroys bacteria by inactivating an enzyme that bacteria need to form cell walls. So some bacteria evolved to produce another enzyme that destroys the penicillin before it can destroy them. Other bacteria evolved enzymes that penicillin can't bind to.

One of the cleverest weapons bacteria use against antibiotics is the plasmid, a piece of DNA that is separate from a bacterium's chromosome. The gene for the penicillin-destroying enzyme, for example, is carried on a plasmid. As a result of the widespread use of antibiotics, most of the bacteria that dwell in humans now contain the gene for that enzyme. The beauty of the plasmid—from a bacterium's point of view—is that it can travel from one bacterium to another, and even from one species to another, spreading mutant DNA. Sometimes viruses that live in bacteria carry plasmids from one host to another. A popular way of moving plasmids is through the bacterial version of sex. Bacteria are unisex, but a plasmid-carrying bacterium will catch the eye, so to speak, of a plasmidless bacterium. The two fuse temporarily, during which "embrace" the bacterium with the plasmid makes a copy of it and gives it to the other. The two then go their separate ways, each now the possessor of the plasmid.

Scientists try to keep one step ahead of bacteria by coming up with new antibiotics. But now many bacteria are resistant to more than one antibiotic; scientists call them multiple-drug resistant. Tuberculosis, which public health officials thought they had almost wiped out, has returned with a vengeance. There are even strains that appear to be resistant to all of the available antibiotics.

Some researchers believe it will be necessary to take a different tack against bacteria, such as developing vaccines against them. In the meantime, however, if your doctor prescribes a course of antibiotics, don't stop partway through, even if you're feeling fine.

Though you may have killed off most of the bacteria, there still may be some lurking within you, eager to come back in full force.

> *Does the fact that bacteria can develop resistance to antibiotics support the theory of punctuated equilibrium?*
> (See page 107: Evolution: Gradual or Sudden?)

VERTEBRATE EVOLUTION

When we accuse people of having no backbone, we mean that they are weak-willed, that they lack character. Yet the evolutionary advantage of the backbone was increased flexibility, not strength; it allowed a wider range of movement and set the stage for greater development of the sense organs and the enlargement of the brain.

The flexible backbone developed from the notochord, a rigid skeletal rod. The vertebrates' earliest ancestors, marine creatures with notochords, filtered their food by the use of hairlike cilia. Instead of a cilia filter-feeding system, the earliest vertebrates—who lived more than 450 million years ago—had a more efficient, muscular filter-feeding system. From a muscular filter-feeding system, the next big evolutionary step was the development of moveable jaws. This probably sounds thrilling only to those who have suffered having their jaws wired shut. It was certainly momentous to those fish to be able to take bites out of their food. In fact, they could take bites out of each other!

The next major step, even more important than the development of jaws, was the movement from water to land. This required a host of adaptations, which many would-be land-dwellers probably failed to develop. Some of their amphibian descendants, such as frogs and salamanders, are still very dependent on the water. Since their eggs are soft and prone to drying out, most of them breed in the water. Their moist skin is also vulnerable to drying out.

The body had a whole new set of mechanical problems to face once it had to deal with gravity without the water to help support it. Legs developed from fins and then strengthened. Paleontologist Stephen Jay Gould points out that most fish did not have fins suitable for the transition to legs, but a small group of fish had evolved an unusual fin structure, for their own adaptation purposes, that happened to have the potential to develop into legs.

Reptiles, the next descendants, were more successful land-dwellers. They had dry skin, eggs with protective shells, and more complex lungs and an expandable rib cage. Thus, the reptile was independent of the water except for its need to drink it. The reptiles' main failing as land-dwellers was that they couldn't tolerate great temperature fluctuations. For marine creatures, this was not a problem, as temperature is more constant in the water than on land.

In response to this problem, two groups with temperature-regulating bodies developed—birds and mammals. To conserve heat, birds were insulated by feathers, and mammals by fur. Both groups had vascular means of controlling their internal temperature. One distinct difference between the birds and mammals was their reproductive systems: birds laid eggs, whereas mammals evolved a means of internal incubation. This left mammals mobile during incubation (no nest to sit on). Development of the mammary glands also freed mammals of the need to seek food for their young.

Eventually, the biggest difference between the birds and mammals came to be brain size. Scientists put forth various reasons for the large size of the mammalian, and particularly the primate, brain. Some scientists suggest the evolution of the hand as a contributing factor. Most animals that stood upright, such as the tyrannosaurus and kangaroo, gradually developed shortened forelimbs. The length of the human arm may be evidence that human ancestors lived in trees. Monkeys, for example, run across branches, while the heavier chimps swing from them with their lengthened forelimbs. The various skills that humans learned with their hands may have encouraged brain growth.

Other scientists suggest that the critical factor may have been the fact that humans were able to breed all year round, rather than just when the female was in heat. This meant that instead of always competing for females for sex, theoretically each male could settle down in a stable relationship with one female and work cooperatively with other males. The more complex society that developed contributed to the brain's growth.

Though we like to think that the whole point of evolution was to lead to us, Stephen Jay Gould also likes to point out that it could easily have turned out differently. During their first 100 million years, mammals were small hairy creatures living in hiding from the dinosaurs that dominated Earth. Had the dinosaurs not been wiped out in a mass extinction, things could still be the same.

If birds evolved from reptiles, how did feathers evolve from scales? (See page 132: From Scales to Feathers)

Why did the dinosaurs die out? (See page 115: Why There Couldn't Be a Lone Dinosaur or Two Alive Somewhere)

Are there as many species of vertebrates, such as mammals, as there are of invertebrates, such as insects? (See page 109: Biodiversity)

FROM SCALES TO FEATHERS

Most of us have at least a vague idea of the evolutionary ladder. Perhaps we recall an illustration of a fish crawling out of the water as a reptile or of an apelike mammal taking its first upright step. But the general sequence is only the larger picture. Though most contemporary scientists consider evolution to be as established a fact as, say, gravity or momentum, as in any scientific field, details and gaps remain to be filled in. For an evolutionary biologist, some of the most important detective work to be done is in piecing together how specific physical features of one species evolved into those of another. One such puzzle is that of how the scales of early reptiles evolved into birds' feathers.

Evolution is a process of successful mutations. In every line of plants or animals, new genetic combinations are always appearing. The results, which are random, are called mutations. The definition of evolutionary success is reproduction. The more numerous a line's offspring, the more successful it is in evolutionary terms. Thus, a successful mutation is one that produces the most offspring. Success favors the mutation that has some kind of functional edge over its rivals. Let's say, for example, that a kind of crustacean called the slimedweller feeds off small fish that dwell in slimy coastal areas. A mutation of the feet that gave the slimedweller a better grip while catching prey would give it an advantage, especially during periods of limited supply. More of these slimedwellers would live long enough to reproduce, thus passing on the genes for that mutation. Eventually all slimedwellers might have the gripping feet, and the slimedweller would have evolved into the creature known as the gripperfoot.

Birds most likely evolved from small dinosaurs similar to a birdlike fossil called the *Archaeopteryx*, which means "ancient bird." The rudimentary feathers of the *Archaeopteryx* probably served some nonflying function more successfully than the scales they developed from. There are several theories as to what that function might have been. One theory is that early feathers served as thermal insulation—in other words, the first down coat. Another possibility is that the *Archaeopteryx* used the feathers as a kind of net, to scoop up insects or other small prey.

Whatever the earliest function served by the feathers, the question remains as to how the birds' early ancestors first used them as a means of locomotion—in other words, for primitive flight. One possibility is that the *Archaeopteryx* was a tree-dweller. Its feathers, which originally developed to provide insulation, expanded until they resembled rudimentary wings, and the *Archaeopteryx* began to use them to glide from tree to tree. If the early feathers were used primarily to catch prey, rather than for insulation, the *Archaeopteryx* may have learned to use its feathered forelimbs to become airborne while chasing prey on its hind limbs. Airborne in this case does not mean actually flying. The bony structure of the *Archaeopteryx*'s body was solid, not thin or hollow like a bird's. But the speed gained by rising even a small amount into the air might have given it an advantage over rivals.

If birds could evolve from dinosaurs, then could humans evolve into the hairless, weak-bodied, large-headed alien-like creatures of science fiction? (See page 113: The Future of Human Evolution)

If Archaeopteryx *lived in trees, could its feathers have served as a form of camouflage?* (See page 136: Animal Camouflage)

FUR

Some people may think that the purpose of fur is to make pets cuddly. The primary purpose of fur, however, is to insulate an animal from the cold by trapping a layer of air against the body. Though some early reptiles may have been fur-covered, of modern animals, only mammals have fur. Hairs do not provide warmth individually, but only as a collective covering. Evolutionary scientists believe that hairs originally served some other function and then became adapted as thermal insulation. Solitary hairs may have served to detect touch and pressure, as do the whiskers of many modern animals. Eventually they grew thickly enough to keep the animal warm.

Each hair grows from a follicle beneath the skin. If the follicle is curved, the hair is curly or wavy. The hair itself is a shaft with a central medulla, surrounded by a layer called the cortex, and then the outermost cuticle. The cells of the cuticle overlap each other like tiny scales. The cortex, which takes up most of the shaft, contains the pigment. Hairs grow in cycles; periods of growth alternate with periods of rest. The human growth period is about three years. Except in animals that molt seasonally, hairs are not synchronized in their growth cycles.

The fur of many mammals consists of two layers—long, outer guard hairs, and a denser, soft undercoat. The insulating air is trapped in the undercoat. Because the outer hairs are coated with oil, the animal is effectively waterproof. When the animal gets wet, it just shakes off the drops to dry off. Some animals, such as the platypus and water shrew, have less outer covering. When they get wet, they run through a narrow tunnel to wring themselves dry.

Most fur grows in a single direction, toward the tail. This direction of lay, as it is called, is caused by the angle of the follicles below the skin. Some animals have whorls, a point from which the hair grows outward. Humans have a whorl on top of the head—thus, the most "natural" hairstyle is the bowl-shaped cut. Attached to each follicle is a small piece of muscle, which can contract to make the hair stand erect. When an animal is frightened, the adrenaline it secretes causes these muscles to contract. With its fur standing on end, an animal looks larger and more threatening.

Humans have long killed animals for their fur, but one animal that doesn't have to be killed for its hair is the sheep. Wild sheep have only a thin undercoat, but domestic sheep have been bred to produce a thick fleece. If domestic sheep are not sheared, their coats become matted and catch on branches and other objects. To make life easier on sheep-owners, if not sheep, Australian researchers have succeeded in injecting sheep with a growth factor that interrupts the hair growth cycle. This causes breaks in the hairs, allowing the owners to peel off the wool with their hands.

In recent years, scientists have become very interested in how hair grows. One reason, of course, is the commercial pot of gold that awaits anyone who can find a cure for baldness. But there are other potential rewards. The growth cycle provides a window into how the body signals its cells when to grow and when to stop growing. This has particular bearing on the study of cancer cells, which grow uncontrollably. But its scientific usefulness notwithstanding, many pet owners still think that the real importance of fur is that it makes pets cuddly.

Could feathers have been another evolutionary mechanism for keeping warm? (See page 132: From Scales to Feathers)

ANIMAL CAMOUFLAGE

The effect of camouflage is deception. You might call it a nonviolent form of self-defense, though it is also practiced by predators, to keep prey unaware of their approach. No matter how subtle or resourceful the camouflage technique or behavior, it is an evolutionary adaptation; animals do not decide between looking like a twig or a leaf for the evening, the way a human agonizes over what to wear to a party.

Camouflage cannot be separated from animal behavior. No camouflage will work unless the animal situates itself among the appropriate surroundings. Frequently, camouflage is part of a triangle of defensive behavior that includes stillness and silence. Blending in with one's surroundings is more convincing if the illusion is not broken by movement or sound. Among birds, the males often have more flamboyant plumage, which they use in courtship and to distract predators from nesting females. Among those species of birds in which the male tends the young, the female is usually more colorfully attired.

The most basic form of camouflage is called disruptive coloration. Here the animal's markings are not designed to blend in with anything particular in its environment; the different shadings simply break up the animal's outline, making it less noticeable. Even the zebra's stripes, which to us are so striking, serve to break up the animal's outline from a distance. A variation on this is markings that help different parts of the body to blend in with one another. Some frogs, for example, have what look like random markings on their body and legs. But when the frog freezes in a resting position, the markings form a pattern of continuous stripes that make the outline of the body harder to perceive.

Since the eyes can be a clear giveaway, masking them is often an important part of camouflage. The camouflage can be as simple as the dark band that runs along the side of the badger's head. Sometimes false eyes appear on other parts of the head or body, to distract from the real eyes.

Some animals change color seasonally, or with different stages of the animal's development. Certain moths, for example, hatch as brown caterpillars for feeding off autumn leaves and hibernating on tree branches. When they awake in the spring, they are green, to match the new leaves. Color adaptation can also demonstrate evolution in progress. A well-known example is the peppered moth, whose light-colored wings speckled with dark spots enabled it to thrive in birch trees and on light-colored walls. In Manchester, England, after the Industrial Revolution, a darker variety, which previously had existed only in small numbers, now successfully blended in with the soot-blackened trees and walls.

The aim of camouflage can also be to stand out rather than to hide. A frightened animal may wish to appear more threatening than it actually is. Some moths have large fake eyes on their wings; some caterpillars have fake eyes on their heads, or even entire fake faces on their rear ends.

Perhaps the most remarkable type of camouflage is that in which an animal alters its surroundings to resemble itself. Certain spiders leave small dark clumps in their webs that serve as decoy spiders. If a predator approaches one of the decoys first, the real spider has a chance to escape. There are even larvae that nosh away at surrounding leaves until the leaves look like the larvae themselves.

The only common form of human camouflage is the brown and green pattern used by soldiers on their clothes and vehicles. Because we have the ability to alter our environments and bodies, there's little likelihood that we'll ever develop stripes on our skin or take on the coloration of the trees in our yard.

If we wanted to change the color of our skin, could we transplant pigment cells from other animals? (See page 191: Monkey See, Monkey Be: Xenotransplants)

If a new variety of moth could develop after as short a period of time as the Industrial Revolution, how quickly can evolutionary changes take place? (See page 107: Evolution: Gradual or Sudden?)

Could we transplant camouflage genes into humans? (See page 178: Genetic Engineering)

EYES THAT GLOW IN THE DARK

Do you think your cat notices that your eyes don't glow in the dark?

The process of vision, in both you and your cat, is basically the same. Light passes into the eye and is absorbed by cells of the retina. A photochemical reaction in the retinal cells then converts light energy into electrical nerve impulses in the optic nerve fibers. The pattern of these signals, the visual image, is sent to the brain and eventually reaches the person's or cat's consciousness.

There are two types of retinal cells—cones and rods. The cones, which are concentrated in the center of the eye, give us our day, or photopic, vision. Because each cone triggers its own nerve cell, the brain can differentiate between signals from many closely neighboring cones. This gives our photopic vision great perception of detail and color. The rods, which are more sensitive and numerous, specialize in scotopic, or night, vision. The rods are bunched up in the sense that many rods activate a single optic nerve fiber. This helps the eye detect objects in dim light.

Some animals, such as moles and shrews, live almost totally in darkness and have evolved eyes that can detect little more than shades of light and dark. Animals that hunt both day and night have evolved eyes able to see in both. To make the most of dim light, the cat's eye has a mirrorlike coating at the back of the retina, called the tapetum lucidum, that reflects back any light that is not absorbed by the retina the first time around. Thus, the retina receives a second stimulation from this "recycled" light. But not all of this light is absorbed, either. The leftover light escaping the cat's eyes is what makes them appear to glow in the dark. A cat can see in light about six times dimmer than what humans require.

Behind the retina of the human eye is a dark pigmented tissue called the choroid. When light rays enter the lens at the front of the eye, they are bent in such a way that they are focused. These focused rays hit the retina. In the human eye, any rays that are not absorbed by the retina hit the choroid, where they are absorbed. If these light rays were not absorbed, they would reflect back to the retina. Because they would no longer be focused, they would make the image less sharp.

So there is a price to pay for the ability to catch mice at night and the glamour of eyes that glow when you hide under the bed. Because cats have fewer cone cells than humans, and a tapetum lucidum instead of a choroid, they probably have fuzzier vision than we do. Your eyes may not appear to glow in the dark, but they're sharp enough to read the fine print on your cat's veterinary bills.

Could a cat's tapetum lucidum be transplanted into human eyes so that humans could see better in the dark? (See page 191: Monkey See, Monkey Be: Xenotransplants)

How do the cones in our eyes detect color? (See page 154: Color Vision)

A CALL TO ARMS: STARFISH REGENERATION

Not all starfish have five limbs. Some have four, six, seven, eight, or even more. The basket star's limbs subdivide over and over till they can hardly be counted. Another type of starfish has such short arms that its body is an almost circular pentagon shape. And although we refer to a starfish's arms, they are not appendages in the sense that our arms and legs are. They are really extensions, or lobes, of its body. On the starfish's underside, called the oral surface, the mouth of the central disc extends in grooves along the arms. The arms also contain branches of the starfish's reproductive system and parts of its digestive tract. Though starfish don't see the way humans do, they have a photosensitive spot at the end of each arm that allows them to react to changes in light.

However many arms it has, the starfish can perform remarkable feats of regeneration. Any part of an arm can regenerate, as can sections of the central disc. A single arm attached to a portion of the disc can regrow an entire starfish, though it can take up to a year for this process to be completed—a long time when you consider that the starfish's life span is five years.

Although most starfish come in male and female varieties that produce sperm and eggs, respectively (which are shed into the water, for fertilization), some starfish use regeneration as a means of reproduction. Their central disc splits into two pieces, and each half then regenerates the missing part of the disc and arms. Some oyster fishermen used to unwittingly assist these starfish in their asexual babymaking. Starfish are predators of oysters, as well as of other bivalves and fish. Oyster fishermen used to cut their enemy, the starfish, in two and throw the pieces back in the water—effectively doubling the predators' population!

The reason starfish can regenerate limbs is related to why we can't. They are simple, and we are complex. Amphibians, for example, go through a metamorphosis from a simple immature creature such as a tadpole to a more complex one such as a frog.

Tadpoles can regenerate limbs, but the more complex frogs they turn into cannot. Being little helps, too. Children up to the age of five or so can grow back the last segment of a fingertip—including the fingernail—if it is cut off. This is probably because of the greater nerve density in their fingertips; as their fingers grow larger, the nerves grow farther apart. Scientists don't yet know how the nerves help to promote regrowth.

Some researchers think that as animals evolved, the genes that aid in regeneration were switched off after embryonic development. In humans, only relatively simple structures, such as bone and skin, can regenerate. Regeneration specialists hope that eventually we'll be able to take what we learn about regeneration from lower animals and use it to enable humans to regenerate limbs and organs. But as for splitting in two as a substitute for sexual reproduction. . . .

If we could develop the starfish's ability to regenerate body parts, could we life forever? (See page 193: Why We Don't Live Forever)

ANIMALS AS ARCHITECTS

Though the beaver may be the most famous animal architect, the need for shelter is so basic that many animals have developed ingenious methods of construction. Some of the lowest animals simply form protective cases around themselves. Certain species of amoeba, for example, add grains of sand to their exterior. The most basic shelter separate from an animal's body is a simple hole in the ground or in a hollow tree. Yet some very simple animals construct surprisingly complex nests. The larvae of caddis flies, for example, take pieces of leaves and other material and glue them together with silk secreted from glands on their heads. Termites build huge, multi-chambered structures with elaborate ventilation systems.

Once we start looking at shelters built by animals, we find not only remarkable artifacts, but also long-standing scientific controversy. On one side are those who argue that even the most elaborate animal constructions are solely the result of genetic programming. In other words, no learning or achievement is involved; the animal is simply following the genetic instructions it was born with. On the other side are those who argue that although higher animals are born with the knowledge of how to build, that knowledge provides only the basic blueprint. The animal learns from its elders and consciously responds to its environment; it even shows "aesthetic" preferences and takes pleasure in its achievement. The animal may also have a concept of the finished product in its mind before it starts construction.

One of the most impressive architects is the bowerbird of Australia and New Guinea. The male builds its bower to attract females; the more attractive the bower, the more compelling the male. A bower is a platform beneath two rows of woven sticks that curve together in an arch. Some males use a piece of bark to paint the walls of their bower with crushed fruit or charcoal, repainting daily. They decorate their bower with leaves, feathers, and fresh flowers that they replace daily. Young males work together building relatively primitive bowers, sometimes with help from more experienced males. After their "apprenticeship," they are ready to build their own bowers.

The beaver not only builds shelters, but also alters its environ-ment by building dams to form ponds. Like any engineer, the beaver often has to try different methods. If the water level isn't rising as it is supposed to, for example, a beaver will vary the material it adds to its dam. According to Donald R. Griffin, a biologist who has studied animal behavior extensively, the beaver's ability to respond to emer-gency situations is strong evidence of its consciousness. He recounts the experience of a scientist who was observing a particular beaver colony. At a certain time each evening, the same male would come out to examine the dam and make necessary repairs. One day, van-dals made a large hole in the dam, causing the water to rush out. The scientist and her companion placed large stones upstream of the dam to try to limit the flow. When the male came out for his evening inspection, he responded to the threat immediately. When placing branches on top of the rocks didn't help, he and three other beavers who joined him used mud and vegetation to plug the spaces between the rocks. Eventually, the water stabilized, though at a level lower than the original pond, and the beavers returned to their lodge. The next evening when the male appeared, he removed a stick from the lodge and took it to the dam to continue the repairs. According to Griffin, this shows that the beaver remembered the previous day's damage and the need to rectify it.

Beavers appear to exhibit memory and intelligence. Are there other animals that exhibit similar traits? (See page 144: Octopus IQ)

Octopus IQ

In the simplistic picture of evolution—the one where fish crawl out of the sea, eventually to stand upright and walk as apes and then humans—we assume that the closer the creatures are to human, the higher the intelligence. To a large extent, this is so; you are no doubt smarter than the tuna in the salad you had for lunch ever was. But a notable exception is the octopus. The octopus belongs to the group of marine animals called cephalopoda, which also includes the squid and cuttlefish. Though all cephalids have well-developed nervous systems that culminate in a brain and a large, complex eye, the octopus seems to be the smartest of the bunch—smarter than the lowest vertebrates, which disrupts our simplistic chart.

Because scientists can't administer IQ tests to animals—and they're of dubious value even with humans—to evaluate an animal's intelligence, scientists usually focus on the two related abilities of learning and memory (learning a task is much more meaningful if you can remember it). Octopuses show distinct strengths in both areas.

Most invertebrates have to experience a situation dozens of times before they make the associations that we would call "learning by experience." Octopuses, however, can usually make such associations the first time, an ability commonly thought to be exclusive to birds and mammals. A good example is an experiment conducted by two British scientists, B. B. Boycott and J. Z. Young, at a famous zoological center in Naples, Italy. The researchers placed a few bricks in an octopus's tank. Octopuses like to hide away in corners and crannies, and this one immediately went behind the bricks. When the researchers dangled a crab at the other end of the tank, the octopus immediately came and captured the crab, and then returned to its bricks. Next, the researchers dangled a crab with an electrified plate behind it. When the octopus grabbed the crab, it received a shock. The octopus became more cautious about capturing the crab, and after a few more trials, would approach the crab only if there was no plate behind it. In similar experiments, where an action resulted in either food or an electric shock, both sighted and blind octopuses were able to distinguish by touch objects that were barely distinguishable visually. Visually, octopuses can also be taught to distinguish geometric shapes, such as squares of different sizes, horizontal and vertical rectangles, and black and white circles.

Recently, scientists at the Naples zoological station found to most of their colleagues' surprise that octopuses can learn by observing their peers. The scientists placed two balls, one red and one white, in a tank. They then trained one group of octopuses to attack the red ball and another group to attack the white ball. The two groups of octopuses then demonstrated their new skill to untrained octopuses. Those octopuses that had viewed the red-ball-attackers attacked the red ball, and those that had viewed the white-ball-attackers attacked the white. Until then, most scientists had thought that only humans and other vertebrates could learn by watching their own species—a process considered to be preliminary to conceptual thinking.

Octopuses can remember what they have learned for at least several days. In one experiment, a researcher offered some large oysters to several hungry octopuses. They tried for several hours, without success, to open the oysters. When the researcher offered them oysters again a week later, they knew not to waste their time on them.

Octopuses, probably the smartest of the invertebrates, appear to be more intelligent than the lower vertebrates. Perhaps one reason it took so long to dispel the myth of the steady march out of the water and to the library is that octopuses are—not to mince words—weird looking. Certainly dolphins, which have been recognized by science as intelligent, don't resemble humans either. But it is easier for us to identify with their gentle "smiling" faces than with a creature whose body consists basically of a big head and eight long arms. If nothing else, the octopus can teach us not to judge by appearances.

How does memory work? (See page 195: Memory)

Can dolphins "teach" each other? (See page 201: Dolphins and Language)

BEES' DANCES

The honeybee and bumblebee are among the most social species of bees. This doesn't mean that they go to bee discos or dinner parties. It does mean that instead of having a strict division of labor, like less social bees, they take on different house chores, such as cleaning the nest, feeding the larvae, and going out for food. Most important, they have evolved a very sophisticated form of dance-based communication.

When a forager bee, as she is called (they are all female), discovers a good food source, she returns to the hive and uses dance to tell her hivemates the location of the food. Because she already carries the scent on her body, she doesn't need to tell the other bees what blossom to look for. Her dance, which she performs on the vertical wall of the honeycomb inside the hive, is very precise and highly flexible. If the food is relatively close to the hive, she does what is called a round dance, making alternate circles to the left and right. The more abundant the food source is, the more vigorous is her dance. Because the food is close by, she doesn't have to indicate direction.

When the food is farther from the hive, she does a more complex dance called the waggle dance. In this dance, she makes a kind of squashed figure-8. When she is moving through the straight portion that separates the two curves, she waggles her abdomen from side to side, then circles back again, alternating curves, to the beginning of the straight portion. The bee communicates a tremendous amount of information through her waggles. As with the round dance, her scent and level of excitement indicate the type and abundance of the food. She indicates distance by the tempo, or number per unit of time, of her runs. Rather than simply indicate the actual distance, apparently she indicates how much energy is needed to reach the food. If the food is uphill from the hive, for example, she indicates that it is farther than the actual distance.

Perhaps the most subtle aspect of the waggle dance is the way in which the bee indicates the direction of the food source from the hive. Through her sense of vision, she knows the direction of the food by its angle to the Sun. In the darkness of the hive's interior, she transposes this to her sense of gravity. With a vertical axis on the face of the honeycomb representing the direction of the Sun, she shifts the axis of her waggle runs to represent the angle of the food from the Sun. Thus, if the food lies thirty degrees to the left of the Sun, her waggle runs will be thirty degrees to the left of the vertical axis.

Like people, bees have dialects. You call cold fizzy drinks soda; your cousin calls them pop. The Austrian honeybee begins the waggle dance when the food is about 275 feet from the hive, the Italian honeybee when it is about 120 feet. The Italian uses a slower tempo than the Austrian to indicate a particular distance. Like people from different regions of a country, the Austrian and Italian honeybee can understand each other, but with minor instances of miscommunication. An Austrian honeybee trying to follow an Italian's directions, for example, will search for the food too far from the hive.

The bee's dance language was discovered in the 1940s by the Austrian zoologist Karl von Frisch, who won a Nobel prize in 1973 for his work. A few scientists claim that although humans can interpret the dances as location indicators, bees rely only on the dancer's scent. Though most bee researchers reject such an extreme claim, they acknowledge the probable role of the bees' other senses, including smell and sound. In 1989, von Frisch's conclusions received strong confirmation by researchers who developed a robot bee that successfully led bees to food sources.

The bee's dance language is tremendously complex, yet the bee's brain is only the size of a grass seed! To appreciate the honeybee's accomplishment, next time you're at a party, try using a dance to tell a late arrival where to find the chips and onion dip.

Bees don't use maps, but people do; how do maps work? (See page 206: Maps)

Do other animal species communicate? (See page 203: Chimpanzees and Language; and page 201: Dolphins and Language)

VIRUSES

A virus is the ultimate bad guest. It tries to convert its host into itself, sometimes killing its host in the process. This unwelcome entity, able to kill a person or animal in less than a day, yet totally dependent on others for its survival, is neither living nor nonliving. You are an obvious example of a living object; a rock is an obvious example of a nonliving object. But there is a spectrum between living and nonliving, and viruses fall somewhere in the middle. The basic characteristic of life is the ability to replicate, or reproduce. A virus does reproduce, sometimes at an alarming rate, but—and this is a big but—without its host, it reverts to an inert crystal.

Viruses are very tiny; millions can fit on the *o* in the word *on*. This is much tinier than the bacteria with which some people confuse them. (Bacteria are a group of single-celled organisms, not all of which are parasitic. Some bacteria cause diseases and infections in people; others cause garbage to decompose.) A virus consists of several strands of genetic material, usually RNA, surrounded by a coat of protein. When a virus invades a host cell, its genetic material fools the cell into accepting it. The virus then uses the cell's enzymes and other resources to duplicate itself. After the virus duplicates itself a number of times, the host cell dies, and the now more numerous viruses go on to invade more cells. Some viruses, called retroviruses, are even more devious. These retroviruses contain an enzyme that converts the virus's RNA to DNA, which is then inserted into the DNA of the host's cell. Thereafter, whenever the infected host cells duplicate, they also duplicate the virus. The HIV responsible for AIDS is one such virus.

The battle against AIDS is particularly trying, not only because of the tremendous human tragedy, but because the virus apparently replicates so fast that it constantly produces new mutants. Any successful treatment or vaccine must work against the many mutant forms of the virus. Though carriers can remain symptom-free for years, researchers think that from the time of infection, the virus reproduces rapidly. At first, the body's T4 cells, which fight the virus, keep up with the virus's rate of reproduction. Eventually, of course, the virus wins.

Antibiotics, which successfully combat many bacteria, are ineffective against viruses. The best strategy against a virus is to develop a vaccine against it. A vaccine uses nonvirulent forms of a virus so that the vaccinated person develops antibodies to the virus. Two phenomenally successful vaccines have been the polio and smallpox vaccines. In 1980, the World Health Organization declared that the battle against smallpox had been won. The only remaining samples of smallpox virus were housed—under very high security—at the Centers for Disease Control and Prevention, in Atlanta, Georgia, and at the Institute for Viral Preparations, in Moscow. The samples were scheduled for destruction in June 1995, but heated debate caused their destruction to be postponed. Those arguing for destruction cited the unacceptable risk of the virus's escaping and causing a worldwide epidemic, even more dangerous now that people are not routinely vaccinated against smallpox. Those who argued against destruction said that scientists still have much to learn from the virus that cannot be learned from noninfectious cloned versions. Nonetheless, in May 1996, the World Health Organization set a new date of June 30, 1999, for destruction of the virus.

The story of viruses is not all illness and tragedy. Scientists are working on ways to use viruses to fight ailments such as cystic fibrosis, epilepsy, and cancer. The strategy is to insert healthy genetic material into viruses and then send these "couriers" to host cells to replace the defective genetic material. Agricultural scientists are experimenting with the use of viruses as "natural" pesticides. In Louisiana, for example, they used a virus to control the velvet bean caterpillar population in soybean fields.

Of course, viruses can be terrifying. We hear about fatal outbreaks of exotic viruses on other continents. But even as we do so, we are probably harboring and fighting off more mundane viruses without thinking twice—or even once—about it.

Could we genetically engineer genes that would be resistant to viruses such as HIV? (See page 178: Genetic Engineering)

CHAPTER EIGHT: **Senses and Perception**

THE SENSES

We are constantly taking in information about the world around us. But we don't simply record the information like some mere instrument that graphs data. We experience the information as sensations; quite literally, we "make sense" of the world. Our five sense organs—eyes, ears, nose, tongue, and skin—receive various kinds of energy from our surroundings. They convert this energy into nerve impulses that travel to the brain, which interprets the impulses and registers them in our consciousness. Even the most abstract information, say, a scientific formula, must go through at least one of the senses to make it to the brain.

Even though the nerve impulses from the different sense organs are similar to each other, the brain can distinguish them by their routes through the nervous system and their destination in the brain. Nerve impulses from the eye will cause a visual sensation even if the stimulus is something other than light. A blow to the eye, for example, will cause a person to see the proverbial "stars."

When light enters the eye, the cornea and aqueous humor focus it first, and then the lens does the fine focusing. The retina, which contains rods and cones, translates the image into nerve impulses. The cones are sensitive to color, and the rods, which enable us to see in dim light, see form and motion. Once the nerve impulses travel to the brain, the brain takes over and organizes them into a visual image that corresponds to the external world. For example, our visual perception of a tiny building includes the assumption that its smallness is a result of distance, not true size. Most scientists think that this aspect of visual perception is learned. When people who are blind from birth acquire sight, for example, they have difficulty in interpreting the visual information they receive.

The richness of the world of sound is due to the wide range of frequencies of sound waves we perceive. The outer ear directs the sound waves, which are basically vibrations, to the eardrum, which vibrates in response. The waves are then transmitted to three small bones of the inner ear. The cochlea, a fluid-filled coil, then transmits the waves as nerve impulses to the brain. As with vision, the brain interprets the nerve impulses into a meaningful whole. With a form of brain damage called auditory agnosia, the nerve impulses reach the brain, but they make no sense to the hearer. The inner ear also contains three fluid-filled canals that help us maintain our balance (when we stop spinning suddenly, the continued swirling of these fluids makes us dizzy).

Both the tongue and nose respond to chemical substances. The tongue is our organ of taste. Scientists usually divide tastes into four basic groups: sweet, salt, sour, and bitter. The taste buds detect chemical substances and then transmit nerve impulses to the brain. Nerves in the front of the tongue send information to the brain about temperature and touch. The brain combines this information with the information from the taste buds. This is one reason, for example, why hot food tastes different from cold food.

It is hard to separate taste from smell, as anyone who has eaten a meal while suffering from a stuffed nose will tell you. Our sense of smell is much more sensitive than our sense of taste. Smell is also very important to memory; just a whiff of a particular smell can often call up entire scenes. The receptors for smell are located behind and slightly above the bridge of the nose. These trigger the nerve impulses to the brain. Although our sense of smell is certainly inferior to that of a dog, it is still quite impressive. A person with a good sense of smell can perceive as many as ten thousand different odors.

Through our skin, we perceive pain, temperature, and pressure—which includes many subtle nuances of touch, from the tickle of a feather to somebody walking on our back. The sense of touch acts as a warning signal to the body. When a hand touches a hot saucepan, for example, the signal goes to the spinal cord and then, bypassing the brain, goes directly to the arm muscles, which pull back the hand in a reflex motion. Some sensations don't warrant more than momentary attention from the brain. The sensations of our clothes touching our skin, for example, fade from our consciousness within moments after we get dressed.

In a fundamental way, we are our senses, for all the elaborate learning, memory, and creativity that our brains engage in are based on our sensory perception of the world. We are also alone with our senses; we can never know if the red coat we see is the same shade of red someone else sees, or what the song we hear on a radio sounds like to someone else.

What is color and how do we see it? (See page 154: Color Vision)

In dreams there are no actual wavelengths of light or sound waves or other sensory input, yet in our dreams we feels as if we're seeing, hearing, tasting, etc. How do we perceive our dreaming "reality"? (See page 185: Sleeping and Dreaming)

How does the brain convert sensory nerve impulses into thoughts? (See page 59: The Chemistry of the Brain)

COLOR VISION

Most of us take the colors of the things around us for granted. Redness is part of an apple's "appleness," just as yellow is part of a lemon's "lemonness." Yet at night the apple and lemon look gray, without losing any of their appleness or lemonness. Although an object's color is a function of how its surface reflects light, color vision is a combination of the physics of light, the structure of the eye, the function of the brain, and psychological factors. When light hits an object, the surface of the object absorbs some wavelengths and reflects others. A green leaf, for example, absorbs all of the wavelengths of light except green, which it reflects. When the green wavelength is reflected back to the eye, the eye and the brain do their part in making the leaf appear green. Sometimes we see green even when there isn't a green wavelength present. If yellow and blue wavelengths simultaneously enter our eyes, we see green.

Our eyes are sensitive only to the small portion of the electro-magnetic spectrum that falls between infrared and ultraviolet that we call visible light. Scientists divide visible light into seven basic colors, or hues—red, orange, yellow, green, blue, indigo, and violet. Each color represents a different wavelength of light, with red having the longest wavelength and violet the shortest. Of course, many wavelengths and colors fall between these seven colors, grada-tions between red and orange, orange and yellow, and so on. A person with good color perception can see as many as one hundred different hues. In addition to hue, scientists speak of a color's satu-ration, or degree of hue. The same wavelength of red, for example, will look pink when of low saturation, and deep crimson when of high saturation. The brightness of the light also affects an object's apparent color.

When light rays enter the eye, they first pass through the cornea, aqueous humor, lens, and vitreous humor. These bend the light rays to focus the image. The image then strikes the retina. The retina contains nerve endings that consist of two types of photore-ceptors (light receptors), rods and cones. The rods, which can see in low levels of illumination, see form and motion, but not color. The cones, which perceive color, require high levels of illumination. There are three types of cones, which have peak sensitivities in three different ranges of wavelength. When the image reaches the retina, the rods and cones transmit electrical impulses along the optic nerve to the optic lobe in the brain, where the image registers in the consciousness.

Partial color blindness, which is caused by cones that are either absent or missing their light-sensitive pigment, is fairly common. Total color blindness, however, is rare, afflicting perhaps one in five million people. The neurologist and writer Oliver Sacks had a sixty-five-year-old patient, an artist, who had lost all color vision through brain damage suffered in an automobile accident. The man's ability to see contrast and forms, however, became very acute. He eventually switched to painting in black and white. He also became a night person, spending his nights wandering around and going to all-night diners. At night he could enjoy his powerful night vision and not think about the world of color he had lost.

What are infrared and ultraviolet wavelengths? (See page 37: The Electromagnetic Spectrum)

Vision is only one of our senses. What are the others? (See page 151: The Senses)

Would we have to sacrifice the clarity of our color vision if we wanted to be able to see in the dark like cats? (See page 138: Eyes That Glow in the Dark)

THE EYES OF PREDATORY BIRDS

With our senses, we survey and navigate our environment. We've developed eyes that enable us to drive in daylight, to clip coupons, and to watch television; ears that enable us to hear a neighbor's horn honking rudely outside our window; and noses that tell us when dinner is ready. Though other species share our world of visual, auditory, and olfactory stimuli, they've developed different capabilities to enable them to navigate their environment, feed themselves, and protect themselves from their enemies.

The sense of sight varies enormously among animals. At one extreme are moles and shrews, which have eyes that detect little more than light and dark. At the other extreme are the predatory birds, with the keenest vision of all animals. Experiments with three species of owls, for example, have shown that they are able to see dead prey six feet away under illumination levels only one-hundredth to one-tenth that required by humans to see. Estimates of the visual acuity of some hawks and eagles range from four to eight times that of humans.

Birds' eyes are large in proportion to their heads. The eyes of some hawks and owls are larger than human eyes. One advantage of larger eyes is that there is more area to accommodate the visual cells. The fovea, a shallow depression in the center of the retina, contains the largest concentration of cones, the visual cells that perceive color and sharp images. The density of cones in the fovea of the hawk is about five times that in the human fovea. Hawks and eagles also have two foveae in each eye—one that points forward and one that points sideways. The sideways-pointing foveae provide a wide range of vision, and the forward-pointing ones provide binocular vision. (In binocular vision, two images from slightly different angles overlap, providing depth perception.) Only animals whose eyes face forward have binocular vision.

Though the eyes of most birds have a flattened shape, the hawk's eye is globular, and the owl's tubular. Both birds have limited peripheral vision—the large, forward-facing eyes of the hawk move very little in the socket, and the eyes of the owl not at all. To compensate, the owl has the famous ability to rotate its head—a full 270 degrees, or about three-quarters of a circle. The hawk, which also rotates its head, has the additional talent of being able to turn its head upside down to get a better look at objects of interest.

Because the owl is nocturnal, its retina has many rod cells, the visual cells that are sensitive to low light. Unlike the cones, each of which has its own nerve cell to transmit signals to the brain, rods operate in small groups, each group sharing a single nerve cell. Thus, the owl sacrifices some resolution to its need to see well in relative darkness. Despite this, the owl is also able to hunt in daylight—in part because its pupil, the hole in the iris through which light enters, has a tremendous range of aperture.

Though all birds, including hawks and owls, have a highly refined sense of hearing, predatory birds generally rely mostly on sight for hunting. The expressions *hawk eyes* and *eagle-eyed* are clearly based on fact.

> *How did birds evolve in the first place?* (See page 132: From Scales to Feathers)
>
> *How do cats see so well in the dark?* (See page 138: Eyes That Glow in the Dark)
>
> *Owls can see in very low light; can some animals "see" in total darkness?* (See page 162: Dolphin Echolocation)

INFRARED RADIATION✓

When we see an object, say, a book or a bird, what we really see is the image created by the light reflected off that book or bird. The only exceptions are objects, such as fireflies, light bulbs, and the Sun, that emit their own light. When the light hits the retina of our eye, it generates nerve impulses, which then travel through the optic nerve to the brain, where the image of the book or cat is formed. Though scientists refer to the entire range of electromagnetic radiation as light, our retina is sensitive only to the small portion of the electromagnetic spectrum that we call visible light. Some snakes have eyes that are sensitive to infrared radiation, the type of radiation that falls just below visible light on the electromagnetic spectrum. What would the world look like if our eyes were sensitive to infrared radiation instead of to visible light?

Infrared radiation is basically heat. We use it in electric space heaters, and in the lights that keep food warm in fast-food chains. Virtually all objects on Earth emit infrared radiation. When we view the world in visible light, we cannot see an object that is blocked by another object, because the closer object blocks any visible light reflected from the rear object. In an "infrared world," if the rear object were much hotter than the front object, it would be visible to us. When you first put on your clothes, until your body heat had warmed them, your body would be visible to the world. And you'd have to be careful when trying on clothes in department stores, for other customers would be able to see you through the dressing room curtains.

If our eyes were sensitive to infrared radiation, they would be most sensitive to the wavelengths closest to those of red in visible light. Scientists call this near-infrared radiation. But even to see near-infrared radiation, our eyes would have to be five to ten times larger than they are—a real boon for optometrists and opticians! Because we're warm-blooded, our body heat would make it difficult to detect temperatures of similar degree—our eyes would probably have to either be insulated or protrude from our head on stalks.

Like visible light, infrared radiation is spread over a spectrum of wavelengths. Presumably, each wavelength would have a different "color." But what those colors would look like, it is impossible to say. Just as people who are blind from birth don't know what colors look like, we aren't able to imagine colors that aren't a combination of those we already know.

Though we cannot "see" infrared radiation, we have instruments that make images similar to photographic images by detecting infrared radiation. Weather satellites use infrared imaging devices to detect temperature differences between Earth and high-altitude clouds. Medical workers use infrared imagers to diagnose breast tumors, because the tumor is usually warmer than surrounding tissue. Indeed, we naturally extend our use of the word *see* to include infrared detection when we place our hand on someone's forehead and say, "Let me see if you have a fever."

What other kinds of radiation are on the electromagnetic spectrum? (See page 37: The Electromagnetic Spectrum)

What is on the other (higher) side of visible light? (See page 160: Ultraviolet Radiation)

ULTRAVIOLET RADIATION

What is all around us, yet invisible; makes us feel good, yet is dangerous to our health? The answer is ultraviolet radiation. Though ultraviolet radiation long seemed innocuous, especially to sun worshippers, we now know that it causes skin cancer. Visible light, which we see by, occupies only a small portion of the electromagnetic spectrum. The spectrum of visible light goes from red to violet. Below red, the electromagnetic spectrum moves into infrared radiation, with its longer wavelengths and lower energy. Above the violet end of visible light, the electromagnetic spectrum moves into ultraviolet radiation, with its shorter wavelengths and higher energy. As the electromagnetic spectrum extends into higher energy levels, the radiation becomes more dangerous.

Scientists divide ultraviolet radiation by wavelengths (and thus energy) into near, far, and extreme ultraviolet. Only near ultraviolet radiation reaches Earth, as the ozone layer of the atmosphere absorbs the rest. (The thinning of the ozone layer has raised the specter of dangerously rising levels of ultraviolet radiation reaching Earth.)

Although our eyes are not sensitive to ultraviolet rays, those of butterflies and some other insects, including honeybees, are. Since flower petals reflect ultraviolet rays, bees with ultraviolet-sensitive eyes would have an obvious survival advantage over other bees. By taking photographs with ultraviolet-sensitive film, we can get an idea of what the bees' world looks like. These photographs show flowers that to our eyes appear to be a uniform color, but to bee's eyes have a darkened center, close to the nectar. Any surface that appears to be a single color in visible light might contain patterns visible only to ultraviolet-sensitive eyes. A hot fashion item might be plain colored jackets that display bright plaids in photographs taken with ultraviolet-sensitive film.

Ultraviolet radiation has many practical uses (we won't consider tanning to be a practical use). One of the commonest is in fluorescent light fixtures. A glass tube with an electrode at either end is filled with a mixture of argon gas and mercury vapor. The inside of the tube is coated with a powder such as phosphor. As electricity passes through the gas, the mercury emits ultraviolet radiation; the phosphor then absorbs the ultraviolet radiation and re-emits it as visible light. Fluorescent lights are much more energy-efficient than incandescent ones, which convert a lot of the electricity into heat instead of light.

The sunlamp is basically an ultraviolet lamp. Though it could do its work (and damage) in the dark, manufacturers usually add infrared radiation (for heat) and visible light, to complete the illusion of natural sunlight.

Most of the light emitted by hot, young stars is in the ultraviolet range. Measurements of ultraviolet radiation provide astronomers with important information on planets, stars, galaxies, and interstellar gas and dust, and can help us to understand the evolution of the stars and galaxies.

An entertaining application of ultraviolet radiation is in "glow-in-the dark" or fluorescent, materials. Basically, a fluorescent material absorbs radiation at one wavelength and emits it at another wavelength. Theaters sometimes use so-called black lights, which are ultraviolet lamps. To the audience, no illumination is visible, yet materials on the stage appear to glow in strange colors. If you shine an ultraviolet lamp in an otherwise darkened room, certain objects that contain fluorescent materials will glow. These might include your clothes, as many detergents contain fluorescent dyes to make white clothes look "really white."

Could life-forms on Earth evolve in response to suddenly increased levels of ultraviolet radiation due to the thinning ozone layer? (See page 107: Evolution: Gradual or Sudden?)

Dolphin Echolocation

Most of us feel comfortable navigating in the dark only in the most familiar of circumstances, such as when we pad through the house in our slippers in the middle of the night. Otherwise, we make our way through the dark only hesitantly. People caught in a large building during a power failure have to struggle against panic as much as the dark. But some animals have a nonvisual way to survey their surroundings, find prey, and keep safe. Among these are the bat, the only mammal that flies, and the dolphin (also a mammal).

A dolphin's underwater surroundings are dark and murky. To "see" its environment, the dolphin uses echolocation, also called animal sonar. The dolphin emits rapid, high-pitched clicks and then listens for the echoes as the clicks bounce back from whatever they hit. As the dolphin gets closer to an object, the clicks become more rapid. The dolphin also turns its head back and forth by several degrees in a kind of scanning motion. Through echolocation the dolphin can perceive both distance and shape with remarkable accuracy; it can detect a one-inch object at a distance of more than 200 feet in open waters.

Scientists are not sure how the dolphin produces its clicking sounds. The sounds may pass from one nasal sac to the other; they may also involve fatty tissue near the top of the head. Oddly enough, the dolphin's ear region is not particularly sensitive; the lower region of the jaw is much more sensitive to sound.

The dolphin's skills at echolocation are not at the expense of its other senses. In fact, scientists think that echolocation is closely aligned with the dolphin's visual skills. Researchers in Hawaii placed an echo-free box with an abstractly shaped object in it in a dolphin's tank. The dolphin needed only a few seconds to investigate the shape of the object through sonar. When the researchers then suspended two objects, one similar to the object in the box, above the dolphin's tank, she immediately touched the correct shape. She was able to do this repeatedly and could also reverse the procedure—

view an object and then identify the shape when two enclosed boxes were placed in her tank.

Can humans learn to echolocate? Many blind people already do, though far less precisely than dolphins. Listening to the echoes of their own footsteps or of their clicking fingers, they can detect an object at about six feet away. Because they experience the phenomenon as a sensation in their facial region, it is called facial vision. Of course, echolocation is only one of many acoustic ways in which blind persons perceive and navigate their environment.

Though there is a great deal scientists don't yet know about echolocation, in dolphins or in humans, significant differences in the structure of the ear and brain probably account for the dolphin's vastly greater skills. So as you make your way down to the kitchen in the middle of the night, don't expect to be able to click your fingers to find that leftover piece of pie.

If dolphins can recognize abstract shapes, can they also use language? (See page 201: Dolphins and Language)

In addition to echolocation, are there any other ways that blind people "see" things in their environment? (See page 171: Blindsight)

Echolocation helps some animals find their way. What other methods have animals developed to prevent getting lost? (See page 164: Internal Compasses)

INTERNAL COMPASSES

Many migratory animals travel extremely long distances each year, sometimes hundreds or thousands of miles. Magnetoreceptors, as internal compasses are called, are only one aid animals use to find their way back home. Many animals have an internal sun compass, which enables them to tell direction from the position of the Sun. This biological compass compensates for the changing of the Sun's position throughout the day. Only birds, which migrate at night, have a star compass. This compass, which does not require an internal clock, is oriented solely by the patterns of the stars; it even works in a planetarium. Other environmental clues used by migrating animals, not properly called compasses, include wind direction, visual landmarks, and scents.

In most animals with a magnetic sense, the internal compass seems to consist of a strongly magnetic mineral called magnetite, located somewhere in the head region. Well-known cases include the homing pigeon, salamander, and certain insects, fish, and rodents. Even some aquatic bacteria contain tiny amounts of magnetite that seem to affect their swimming behavior (don't ask where a bacterium's head is). The range of sensitivity is very narrow; that is, the internal compasses do not seem to detect magnetic fields much weaker or stronger than Earth's. Recent research suggests that in at least some animals, the ability to perceive compass direction may involve light-receptors that pick up information from Earth's magnetic field.

As intriguing as the idea of an internal compass is, it is only half of the mystery of how animals find their way home. If you get lost in the woods, using your compass to determine which way is north solves only half your problem. You have to know where you want to go; in other words, you need an internal map in order to make navigational decisions. Scientists define navigation as the ability to find one's way from an unfamiliar location without any direct sensory clues from the destination, such as scent. When an animal simply retraces its steps, scientists do not consider that to be true navigation. Of all the animals studied, birds show the clearest evidence of being true navigators; yet scientists still do not understand how a bird knows which direction home is.

Researchers on magnetoreception conduct what they call displacement/release experiments. They move the subjects—whether pigeons or human beings—to distant locations and then release them. In one case, researchers applied artificial magnetic fields to pigeons before releasing them, expecting that the fields would disorient the birds. To their surprise, these pigeons showed even more accurate orientation. This may be because the applied magnetic field helped to align the magnetic particles of the pigeons' internal compass. The results of experiments involving humans wearing magnetic bars on their foreheads have been inconclusive. So, until more definitive results are in, you might do well to carry a compass in your pocket.

Do animals have a way to tell what time it is? (See page 182: You Gotta Have [Circadian] Rhythm: Biological Time)

PAIN

Pain is useful. Besides giving us something to complain about, it acts as a warning system, letting us know when something is wrong with our body and in need of attention. Loss of sensation, and the attendant inability to feel pain, can be dangerous. One reason people with diabetes sometimes develop infections in their feet, for example, is that nerve damage diminishes their ability to feel pain, leaving them unaware of blisters and cuts. Some nerve defects cause the opposite problem, the transmission of false information, as when people experience pain where amputated limbs used to be— so-called phantom limbs.

Like vision and hearing, pain is a sensation that travels from receptor cells, through the nervous system, to the brain. With vision and hearing, however, the brain basically registers information received from external stimuli. With pain, the brain registers information—say, that a drop of boiling water has fallen on the hand— and then produces an intense, unpleasant sensation. To complicate matters even further, sensations and tolerance of pain are closely bound up with psychological factors such as memory, expectation, and anxiety.

The pathways that carry signals about pain and temperature are not yet as clearly understood as other sensory pathways. Sensations of pain are picked up by nerve endings; they then travel along nerve fibers to the spinal cord. One type of fiber, called A-delta fibers, transmits signals about the fast, intense pain that accompanies a sudden injury. The other type of fiber, C fibers, transmits signals about dull, insistent pain. While the signals are traveling through the spinal cord, they can be influenced by other stimuli such as touch and pressure. That is why rubbing a sore area can help relieve pain. Signals traveling up the spinal cord can also be affected by other signals traveling down from the brain. If a person's attention is drawn to something very interesting or important, for example, pain signals may be repressed. A familiar example is when an injured athlete continues to play, temporarily indifferent to what otherwise would be insufferable pain.

From the spinal cord, the signals travel up two pathways to the brain. The main path leads to the rear of the thalamus in the brain-stem, from which the pain radiates out to other parts of the cerebral cortex. The other pathway sends the signals to parts of the brain involved with motivation and behavior. Scientists think that nar-cotic medications may affect this pathway, since patients who take them continue to feel painful sensations but are less affected by them. The brain itself produces substances called endorphins that have a similar effect. Endorphins travel downward from the brain to the spinal cord, where they help to alleviate pain. In some experi-ments, researchers gave subjects placebos, saying that they were pain relievers. Those subjects whose pain was alleviated had ele-vated levels of endorphins in their blood. In a sense, the placebos really did "work."

Many people suffer from chronic pain—such as headaches and lower-back pain—whose cause cannot be pinpointed. In recent years, medical professionals have begun to specialize in what they call "pain management." Their approach includes such techniques as biofeedback, hypnosis, and visual imagery. Perhaps someday we'll be able to simply will our endorphins into action.

How does the brain process the signals it receives from the nerves? (See page 59: The Chemistry of the Brain)

If we can "learn" to control pain, can we also learn to control other biological processes? (See page 187: Respiratory Control)

PHANTOM LIMBS

Most of us never think about how we know where our arms or legs are. If we did, we might end up like the centipede that thought about walking and ended up unable to move. Knowledge that we unconsciously depend on is called implicit knowledge, and the body of implicit knowledge that tells us about the spatial properties and conditions of our body is called our body schema. Some people who have an inaccurate body schema have to grapple on a daily basis with the physical sensations of limbs that are no longer physically present, while others, less common, lack the sensation of owning one of their own limbs. The latter may try to push their leg out of bed, not recognizing it as their own. Both conditions have to do with the information the brain relays about the body and where it is in space.

The brain is an extension of the spine—an extremely complicated extension. Cells called receptors constantly send messages to the brain about conditions in the outside world and in the body itself. Exteroceptors receive input from the external world through nerve fibers in our skin and sensory organs, such as the eyes and ears. Interoceptors receive input about the body itself through nerve fibers in muscles, tendons, joints, and organs such as the bladder. Other receptors include thermoreceptors, which report on warmth and cold, and chemoreceptors, which report on sensations such as smell and taste.

The cells of the nervous system, which are called neurons, pass messages from one to another in the form of electrical impulses. Certain neuron pathways go from the outer parts of the body to the spine and brain, and other pathways go in the reverse direction from the brain to the spine, and then to the muscles and glands. Between every neuron and the next is a tiny gap called a synapse. Impulses passed from one neuron to another must jump over the synapse between them. Scientists call this the firing of neurons. They now think that disturbances in the firing of neurons are the cause of misleading messages about phantom limbs.

Though some people born without an arm or leg have a phantom limb, phantom limbs are most common with amputees. Amputees experience their phantom limbs as intensely as they do their other limbs. They may have an urge to step onto a phantom foot, or to reach for a door with a phantom arm. Sometimes they sense their phantom limbs as stuck in unusual positions and have to make allowances for them when moving through narrow spaces, for example, or sleeping.

Phantom limbs can experience a range of sensations, including temperature changes, pressure, and itching (scratching in the place the limb would be sometimes relieves the itching). Unfortunately, up to 70 percent of people with phantom limbs suffer from pain, which is sometimes severe. Doctors used to cut the nerve cells at the end of the stump, to prevent them from sending impulses to the brain. They also cut nerves along the spinal cord itself and removed areas of the brain that receive the impulses from the stump. Although these techniques often provide relief, the pain often returns, sometimes even years later. And none of these techniques removes the phantom limb itself.

A prominent pain researcher named Ronald Melzack has suggested that the brain contains a network of neurons that sends out a pattern of impulses telling the body that it is intact and that it belongs to itself. Melzack calls this pattern a neurosignature. The neurosignature is very complex and involves a number of areas of the brain. When the neurosignature operates in the absence of a limb, the person experiences a phantom limb. Melzack takes the fact that people born without limbs sometimes have phantom limbs to suggest that the neurosignature is prewired in the brain. Thus, the brain does not just receive messages from the body's receptors, but actively sends out its own messages about what constitutes the body. This raises questions about self-definition. If a person's neurosignature and the physical reality disagree, which one is "really" the person?

Could scientists eventually devise a method of growing new human limbs? (See page 140: A Call to Arms: Starfish Regeneration)

Does the brain itself function in the same way that the nerves do? (See page 59: The Chemistry of the Brain)

BLINDSIGHT

We often say, "I'll believe it when I see it." But some people who are blind due to brain injury are in a position to say, "I'll believe it when I don't see it." These people, who have what scientists call blindsight, perceive certain visual information with no conscious awareness that they have done so.

Blindsight was discovered by an English psychologist named Lawrence Weiskrantz. He and other scientists didn't learn about blindsight until they changed the questions they asked their subjects. They knew that animals that had had their visual cortex (the part of the brain that receives signals from the eye) removed could often still distinguish between different visual stimuli, such as objects of different shapes or textures. But in studying humans with injuries to the visual cortex, they would ask questions such as, "Can you see the object?" or "Do you know where the object is?" The person, who could not "see" the object, would answer no. Eventually, researchers began to use what they call forced choice methods, borrowed from their animal studies. Instead of asking a person where an object is, they ask whether the object is in location A or location B. The subject, who does not consciously "see" the object, guesses A or B—or thinks that he or she is guessing. A person with blindsight may "guess" correctly virtually all the time. A person with blindsight may also be able to tell whether a line of light is vertical or horizontal, or whether an object is, say, square or triangular in shape.

Until the discovery of blindsight, scientists thought that all of the nerve fibers in the optic nerve at the back of the eye went to the visual cortex. But if this were the case, it would be hard to explain the phenomenon of blindsight. They now know that about 15 percent of the nerve fibers in the optic nerve go to areas other than the visual cortex. The reason more people with damage to the visual cortex don't have blindsight lies in the organization of the brain: other areas of the cerebral cortex are so close to the visual cortex that they are unlikely to escape damage when the visual cortex is injured.

As frequently happens, the study of the abnormal or unusual enlarges our understanding of the normal (just as figuring out why something tastes awful can help you learn to cook). Through studying blindsight, scientists have gained a more detailed understanding of the brain's complexity. The study of blindsight may also help scientists to understand the unconscious processing of other kinds of information, such as sound, touch, and language—which, in turn, may help them to understand why and how certain information does reach our awareness. At that point, philosophers join the fray, for how we know what we know is one of life's great mysteries.

How does the brain process the signals it receives from the optic nerves? (See page 59: The Chemistry of the Brain)

DNA: OUR GENETIC CODE

Imagine that our alphabet were reduced to four letters, and using those four letters, you could write the plans for the construction of all living creatures, from simple bacteria to human beings. To a great extent, that's how things work. Our entire genetic code, which is arranged on strands of deoxyribonucleic acid (DNA), is based on combinations of four letters. The code appears to be the same for all forms of life—yet more evidence that we all originated from the same primordial soup. We still share a lot of our DNA with our close animal relatives. The genetic map of the great ape, for example, is almost identical to that of humans.

The nucleus of every living cell contains threadlike structures called chromosomes. Each species has a distinctive number of chromosomes; humans have two pairs of twenty-three, or forty-six. Chromosomes consist of DNA molecules, which are extremely long polymers (large molecules of repeating units) twisted into DNA's famous double helix structure. If the DNA in a single set of human chromosomes were uncoiled and laid out straight, it would be about six feet long. Segments of the DNA, in turn, contain genes—approximately 100,000 in every human cell. These genes contain the coded information. The system is both incredibly simple, in that all organisms use the same code, and incredibly complex, in the variety of organisms coded and the process by which the code's instructions are carried out.

The four letters of the DNA "alphabet" are the names of four bases—adenine, guanine, cytosine, and thymine (A, G, C, and T). The molecules that contain these "letters" are called nucleotides. Each nucleotide consists of a phosphate, a sugar, and one of the four bases. If the double helix is thought of as a spiral ladder, the phosphates and sugars form the sides of the ladder, and the bases meet in the middle, each pair of bases forming one rung. The bases A and T can pair only with each other, as can G and C, making four possible combinations—AT, TA, GC, and CG. This makes it easy for DNA to replicate during cell division. The two strands of DNA (sides of the ladder) separate, and each unattached base finds its partner among nucleotides in the cell nucleus.

Our bodies are made of proteins, so the genetic code is really a code for protein synthesis. Proteins, which are also polymers, consist of twenty different amino acids; the sequence of amino acids determines the characteristics of the tissue produced—everything from the heart to the kidneys to the eyeballs. DNA has a partner, so to speak, to help it carry out its instructions for protein synthesis. This partner is ribonucleic acid (RNA). RNA is a single strand of the same four bases. First a strand of DNA transfers the code to a strand of RNA called messenger RNA (mRNA). The mRNA then goes to the ribosome, the site of protein synthesis. There, molecules of another type of RNA, called transfer RNA (tRNA), attach themselves to their coded counterparts of the mRNA. The unattached end of each tRNA molecule carries one of the twenty amino acids. Thus, as the tRNA molecules take their places along the mRNA, the amino acids attached to them line up in the proper order to form the desired protein. Viruses, the only living organisms that don't contain DNA, carry their genetic code on the single-stranded RNA.

Molecular biologists speak of a Central Dogma, though Frances Crick—co-discoverer with James Watson of the structure of DNA—says that when he coined the phrase, he didn't really know what the word *dogma* meant. According to the Central Dogma, the information in the genetic code is on a one-way path; that is, it can pass from DNA to RNA, and from RNA to protein, but never in the other direction. Later, scientists realized that the only exception is that information from the RNA in viruses can transfer to the DNA of a virus's host (which is how the virus does its nasty work).

The characteristics that make a person's DNA unique are distinctive enough that scientists speak of our "genetic fingerprint," and courts accept DNA samples as a means of identifying individuals. However, the next time you're at a social gathering where you feel you don't belong, just remember that about 99.9 percent of the DNA is the same in all humans.

If DNA is capable of containing information as complex as the "recipe" for individual organisms, could it be used for computing? (See page 208: DNA Computing)

How do we know what different genes do? (See page 176: The Human Genome Project)

THE HUMAN GENOME PROJECT

In 1990, scientists began a project so ambitious that some refer to it as the Holy Grail. This project is the Human Genome Project, a multinational effort to map the approximately three billion pairs of chemical bases that are the code for the 100,000 genes in our DNA (the word *genome* means a complete set of chromosomes). The motivations extend far beyond the reason for climbing Mount Everest—because it's there. An understanding of genetics underlies most areas of biology, including cell biology, immunology, and neurology. Knowledge of how our genes transmit hereditary characteristics can also foretell how an individual will develop and age, and how species have evolved over the course of life on Earth.

For any given area, a cartographer can create maps of greater and greater detail ranging from a rough sketch to a highly detailed scale map. The mapmakers of the Human Genome Project are following the same path. The first details one notices are those that stand out in some way. We notice a skyscraper among cottages, and vice versa. With genes, it is easiest to pinpoint a mutant gene because it makes a segment stand out from the same segment in other samples. Thus, to a great extent, our earliest genome maps are maps of mutations. Because the mutations we care about the most are those that cause serious diseases, these have gotten the most attention. Scientists have already located the genes for such diseases as hemophilia, cystic fibrosis, and sickle-cell anemia. Locating these genes enables scientists to proceed with their mapping, and it can also offer immediate practical applications.

Before making a precise map, a mapmaker first sketches out relative locations of objects, e.g., place B is between places D and G, and is closer to D than to G. This is what scientists do in genetic linkage mapping: they determine relative locations of various genes, without yet determining their exact location on a chromosome. Mapmakers also use various kinds of markers to help them keep track of where they are. Genetic mapmakers have several types of markers to help them. One such marker uses enzymes to cut the DNA wherever it finds a particular code sequence; thus, you can tell where you are on a strand of DNA by where the enzyme has cut it. When specific genes, such as those that cause the diseases mentioned above, are located, they, too, become markers.

Though the scientists' goal is a complete map of the human genome, with all 100,000 genes identified and located, the process is undertaken as a series of maps, each more detailed than the next. Before the goal is reached, society will have to start looking at the medical and ethical dilemmas that are sure to arise. Is it wise to tell people that they are predisposed to come down with particular diseases if there are no known cures for those diseases? How will we keep insurance companies from abusing the knowledge available from genetic testing? It is certainly true that with knowledge comes responsibility.

What is DNA? (See page 173: DNA: Our Genetic Code)

Once scientists have located a gene, can they change it or replace it to alter the organism? (See page 178: Genetic Engineering)

GENETIC ENGINEERING

Though you will probably never look through a catalogue to pick out the traits you'd like for your offspring, scientists are already using genetic engineering to breed animals that are economically more valuable. Gene splicing—recombinant DNA is its technical name—is a complex technology based on surprisingly simple concepts. One characteristic of DNA makes the most futuristic-sounding scenarios theoretically possible: because the genetic code that makes up DNA is the same for all living organisms, DNA can be transferred from one species to another. In other words, DNA doesn't care whether it is transferred to a caterpillar, a cow, or a human being. Of course, many other factors complicate the picture; you can't just put human DNA into a caterpillar and then teach the caterpillar algebra.

The first step in gene splicing is to cut strands of DNA from different organisms into manageable segments. The cleverest method uses a type of enzyme in bacteria called a restriction enzyme. Restriction enzymes aim for specific sequences of the four-letter code that makes up DNA, allowing scientists to control where the DNA is broken. Segments from the different organisms then bind with each other, creating strands of DNA with genes from both organisms. Scientists can use this technique to insert genes into viruses and plasmids (independent segments of bacteria DNA) that act as carriers of the inserted genes, or to insert genes into eggs and sperm—so far only of plants and nonhuman animals.

Recombinant DNA technology offers both intellectual and practical benefits. Scientists can use the technology to study the workings of genes and the evolutionary history of different species. Scientists are already applying gene-splicing techniques to plants and livestock—not without controversy. A genetically engineered bovine growth hormone enables farmers to get more milk from cows. Though scientists say the hormone is safe because it is structurally different from human hormones, there could be indirect risks. Cows injected with the hormone have a higher incidence of udder infection. Residues of the antibiotics used to treat the cows could end up in the milk, thereby helping bacteria in our bodies develop resistance to antibiotics.

Some people are also concerned about possible consequences of eventual medical applications of genetic engineering, so-called gene therapy. In one type of gene therapy, scientists replace genes in somatic cells, which are cells other than sperm or eggs. Such gene therapy might replace a gene that causes a chronic disease with a "healthy" gene. In the second type of gene therapy, called germ-line therapy, scientists alter the genetic makeup of the sperm and eggs, which affects the offspring. Because the consequences of germ-line therapy affect future generations, many scientists and concerned laypersons are wary about the possible dangers of such a powerful tool. In the worst-case scenario, the technology could be used in a misguided attempt to produce a "master race." Even when applied with the best intentions, germ-line therapy could have unintended biological consequences such as reducing the size of the gene pool. Scientists may also fail to recognize the benefits of otherwise "defective" genes. For example, two copies of a particular gene cause sickle-cell anemia, yet one copy of the same gene protects the carrier from malaria.

The ability to alter our genetic makeup is a very powerful one. Members of families in which diseases such as cystic fibrosis and Huntington's disease are passed from generation to generation, sometimes unknowingly by symptom-free carriers, are among the first to say that the potential benefits are more than worth the risks.

How can viruses be used as carriers of genetic material? (See page 148: Viruses)

How can the bacteria in our bodies develop resistance to anitbiotics? (See page 125: The Growing Resistance of Bacteria to Our Medicines)

TO BE OR NOT TO BE (MALE)

Our genetic code is contained in threadlike structures called chromosomes. Humans have two pairs of twenty-three chromosomes, or forty-six. Twenty-two of the pairs are identical; the last pair, which determines sex, are not. The human female has two X chromosomes (XX), and the human male has an X and a Y (XY). Yet despite this difference, until about the seventh week of pregnancy, an embryo has the potential to be male or female. For an embryo with a Y chromosome to successfully develop into a male, it's a race against the clock to avoid being set off course by the mother's female hormones and turned into a female.

At five weeks, the embryo has a pair of ridges with the capacity to turn into either testes or ovaries, and these ridges contain cells that could become either sperm or egg cells. If the embryo develops rudimentary testes, the testes destroy the developing female organs and cause the male organs to develop. If the embryo has no testes, then the female reproductive system continues to develop. In a sense, the fall-back position is to become female, unless directed otherwise by male tissue. A delay in the development of the male system can cause the testes to become ovaries. Some scientists have suggested that the race to become male before it is too late may influence the male's metabolism throughout his life and account for the fact that males have a shorter life expectancy than females.

Scientists have been searching for a gene that acts as a switch, to turn on the activity that makes a Y-carrying embryo turn into a male. Though the simplicity of such a switch is appealing, some scientists believe that no single gene plays such a role. One complication to the search for such a gene is that about 1 in 20,000 men has two X chromosomes and no Y chromosome. Though some of these men carry some of the genes usually found on the Y chromosome on one of their X chromosomes, about one-third of them do not have those genes at all. Men with two X chromosomes are sterile and have little facial hair (among those races that normally have facial hair). Men who have a Y chromosome but also have an extra X chromosome (XXY) also have problems. This condition, which is called Klinefelter's syndrome, usually causes extra-long limbs, undescended testes, and sterility. In rare cases, men have three, or even four Xs (XXXXY). The more Xs, the more undeveloped are the genitalia. The allotment of X and Y chromosomes has nothing to do with sexual orientation; homosexuals have the same male or female chromosomes as heterosexuals.

Humans tend to be most comfortable when things fit neatly into categories, and sex is no exception. From the beginning, we're taught that everyone is either male or female. Yet as many as 4 percent of humans born are what scientists call "intersexual." Such people may be predominantly one sex with characteristics of the other, or they may be what scientists call a true hermaphrodite. A true hermaphrodite has one ovary and one testis; sometimes the testis and ovary are separate, and sometimes they are grown together. Frequently, one of the organs functions, producing either sperm or eggs. Pseudohermaphrodites have the normal XX or XY chromosomes and either male or female reproductive organs, but they may develop external genitalia of the other sex. Females may develop beards and deep voices.

Medical professionals usually intervene when intersexuals are very young. Parents are counseled to "assign" a sex to their child, and the intersexual undergoes a combination of surgical and/or hormonal treatment. Recently, however, some less orthodox researchers have questioned society's need to deny the presence of intersexuals. If as many as 4 percent of the population are born with bodies that don't fit neatly into one of two categories, then perhaps two categories aren't enough.

How do scientists know which chromosomes determine sex? (See page 176: The Human Genome Project)

YOU GOTTA HAVE (CIRCADIAN) RHYTHM: BIOLOGICAL TIME

Every afternoon at the same time you feel an irresistible urge to nap. It's all in your head—that is, in a bundle of nerve cells deep in the brain. These nerve cells register cyclic signals from the environment, most importantly changes in daylight and darkness, and set off a host of other biological cycles in the body. Your blood pressure, bodily temperature, hormone production, metabolism—even your sensitivity to allergens and environmental toxins—rise and fall in twenty-four-hour cycles.

Circadian (from the Latin *circa*, "about"; *dies*, "day") rhythms are not unique to humans. Almost all animals and plants, even unicellular algae, experience them. This suggests that biological clocks, as they are called, provide an evolutionary advantage; that is, organisms that have them are more likely to survive. One explanation is that a biological clock enables one not only to react to environmental events such as nightfall, but to anticipate them. Let's say there are a limited number of caves in which to take shelter from the cold nights. The animals that could anticipate nightfall would be the first into the caves.

Scientists have known for decades that the body uses a variety of clues to maintain its internal clock; these include light and darkness, diet, physical activity, and social indicators such as clocks, calendars, and social interaction. More recent research suggests that light and darkness may be the most important of these time landmarks, or *zeitgebers* (from German for "time givers").

The first step for scientists was to prove that humans do, indeed, possess internal clocks that function independently of clock radios and roaring garbage trucks outside the bedroom window. These experiments, named *hors du temps* or "out of time" by the French, were first conducted in the 1970s and 1980s. The setup is obvious: place the subjects in isolation and see what kind of wake/sleep cycles they develop. Europeans tended to favor more exotic—and psychologically stressful—locales such as deep underground caves. Scientists in the United States favored hospital isolation chambers visited by teams of technicians.

Basically, what researchers found was that the human clock tended to slow down periodically, the most common cycle being a twenty-five-hour one. In the late 1960s and early 1970s, however, a controversial French researcher named Michel Siffre claimed that after periods of isolation up to three or four months long, some people shift to days as long as forty-eight hours, with about thirty-five hours of waking activity and thirteen hours of sleep. Siffre was a geologist and speleologist (cave scientist) who descended without his clock into an underground glacier he had discovered. He lost total track of time, and the experience caused him to switch to the field of chronobiology, the study of circadian rhythms. His claims about the forty-eight-hour cycle were later confirmed by two prominent U.S. researchers, Elliot Weitzman and Charles Czeisler.

Another effect—and danger—of such out-of-time research is depression and psychological instability. In 1990, a French woman who had been a subject of one of Siffre's underground cave experiments committed suicide. According to her husband, she had never recovered from the debilitating effects of the 105-day isolation.

Knowledge of our internal clock has far greater ramifications than simply convincing college students of the counterproductivity of staying up all night to study. Physicians, for example, now know of the importance of timing medication according to cycles in its effects on the body. The general public has grown aware of the safety dangers posed by having people, such as truck drivers and airline pilots, work under the strain of night shifts and changing shifts. In a study of medical interns, Charles Czeisler found that in a period of a year, more than one-quarter of them had fallen asleep while talking on the phone!

Scientists have begun to experiment with light as a tool for correcting disruptions of the biological clock, from the everyday such as jet lag, to the more critical such as those tentatively associated with clinical depression and even bipolar disease. The technology need not be terribly complicated or expensive—basically, it consists of timed pulses, of one to several hours in length, of very bright light. A man from Bethesda, Maryland, devised a light visor which is worn on the head. Worn during long flights (after a bit of explaining to one's seatmates), it helps to prevent jet lag by resetting the wearer's biological clock.

Do biological clocks change during the life cycle of an organism? (See page 185: Sleeping and Dreaming)

SLEEPING AND DREAMING

There is no record of a human being who did not need to sleep. In the past few decades scientists have learned a lot about the mechanics of sleeping and dreaming, but why we spend about one-third of our time unconscious remains something of a mystery. Although answers range from keeping us out of trouble between meals to strengthening our memory, sleep does seem to serve biological functions. During sleep, for example, output of human growth hormone rises, which in turn increases the rate of protein synthesis.

Our need for sleep changes throughout our life. Newborns sleep the most, around sixteen hours a day. Many teenagers sleep ten or eleven hours a night, and adults around eight hours. Then the need for sleep tapers off; elderly people usually need only six hours. Though some people have gone as long as two weeks without sleep, their waking state is accompanied by paranoia, hallucinations, blurred vision, and failing memory and concentration. Even much milder sleep deprivation affects concentration. A person deprived of sleep will function well on short-term tasks that are extremely interesting, but do poorly on those that are tedious and of longer duration. Not good news for someone trying to drive all night or study around the clock.

The sleep cycle is divided into non–rapid eye movement (NREM) sleep and rapid eye movement (REM) sleep. Scientists determine the stages of the sleep cycle by measuring the brain's electrical activity with a machine called an electroencephalograph, or EEG. NREM sleep is divided into four stages, 1 being the lightest stage of sleep and 4 the deepest (many elderly, however, skip stage 4). Each cycle lasts about ninety minutes. We pass progressively from stage 1 to 4; after stage 4, the sequence reverses back to stage 1, which is then followed by five to fifteen minutes of REM sleep. REM sleep, which in many ways resembles the light sleep of stage 1, is when we dream. Some scientists believe that it is only REM sleep that the body really needs.

People have viewed dreams as everything from random messages fired from the brain's neurons, to wish fulfillment, to the expression of our darkest secrets. Dreams are the result of electrical activity in the brain, but anyone who has awakened appalled or delighted by a dream knows that the activity is not random. Whether there is any real significance to dream content is another question. When researchers wake subjects frequently from REM sleep, many of their dreams turn out to be quite mundane. And even with more tantalizing dreams, researchers are divided as to whether their content should be taken at face value or symbolically. Does a dream about chewing bubble gum just mean that you enjoy exercising your jaw (or annoying others), or does it stand for a pink shirt that your kindergarten classmates teased you about, making you forever insecure about your taste in clothes?

The brain may also have a daily fantasy quotient it needs to fill. In one experiment in which researchers deprived subjects of REM sleep, those subjects who didn't usually fantasize a lot during waking hours began to do so much more. Other researchers have found that when we are in a monotonous environment, our daydreaming follows a cycle similar to REM sleep, with periods of more intense daydreaming occurring every 90–100 minutes.

Environmental factors can influence dream content. A blanket falling to the floor can cause the cold sleeper to dream of being lost in a snowstorm; the sound of a cat banging a kitchen cabinet can turn up in a dream as a marching band. Though some companies sell learning tapes—"Learn to speak French in one month while you sleep!"—most researchers discount the effectiveness of such tapes. But even if you can't claim to be studying when you're still in bed at noon on a Saturday, you can at least claim that getting your fill of REM sleep will keep you attentive while studying.

Why do we sleep every night? (See page 182: You Gotta Have [Circadian] Rhythm: Biological Time)

What is protein synthesis? (See page 173: DNA: Our Genetic Code)

RESPIRATORY CONTROL

Even the least musical of us is a virtual walking rhythm orchestra, as our heart, pulse, and respiration each maintain a steady beat. As with any good music, variations keep the rhythms from becoming monotonous. Like a bored concertgoer, we often ignore the music unless it suddenly gets louder or speeds up. We rarely pay attention to our heartbeat—most of us have far more pressing demands on our time—unless we feel it pounding in our chest from fright or anxiety. We pay attention to our breathing more frequently, usually when short of breath from climbing stairs or running to catch a bus. Joggers pay attention to their breathing as they pace themselves or push themselves beyond their limits.

Although we have no control over the beating of our heart, our breathing gives us the illusion of control. We can make ourselves breathe slowly and deeply, or rapidly and shallowly. Singers learn to control their breathing so they can get the most out of their instrument, their voice. And we can hold our breath—to avoid obnoxious fumes, or for lack of anything better to do—but only for so long. Long before we could die from lack of oxygen, the part of the brain that controls our respiration would override our poor judgment and force us to breathe again.

Scientists call the function of organs like the heart and lungs involuntary. These organs are kept in motion by rhythmical discharges of nerve impulses in the brain. Four centers in the brain stem control respiration. The inspiratory center sends out nerve impulses that cause the muscles of the chest cavity to contract so the lungs can fill; the expiratory center sends out impulses that interrupt those of the inspiratory center, causing the lungs to empty. A third center, the pneumotaxic, acts as a messenger between the inspiratory and respiratory centers, to let the respiratory center know when the peak of inspiration has occurred; thus, the pneumotaxic center controls the rate of breathing. A fourth center, the apneustic, controls the depth of the breaths.

Through respiration, we make a simple trade with the environ-ment: we take in oxygen, which we need to produce energy, and expel carbon dioxide as a waste product. The mechanics are rela-tively simple. The muscles of the thorax and diaphragm cause the chest cavity to expand. Air that has entered through the nose or mouth and traveled down the airways then fills the lungs. When it is time to expel the air, the muscular activity ceases; the ribs fall back into place, and the air is squeezed from the lungs. To meet the body's changing needs, the process must operate on a sensitive feed-back system. If the oxygen level falls too low, or the carbon dioxide level rises too high, the rate or depth of respiration must increase. Because the brain itself is particularly sensitive to loss of oxygen, out of self-interest, as it were, the brain instructs the respiratory system to do something about the problem.

Could we eventually transplant gills from fish into humans so that we could breathe underwater? (See page 191: Monkey See, Monkey Be: Xenotransplants)

In the same way that the brain directs the lungs to breathe to prevent asphyxiation, are there other mechanisms by which the nervous system acts to protect the body? (See page 166: Pain)

USING BACTERIA AS AN INSULIN FACTORY

Bacteria, the one-celled microbes that bring us bronchitis and pneumonia, are so hard to combat because they are able to develop ways to resist our antibiotics. Like a tourist who accidentally walks off with someone else's luggage at the airport, a bacterium will easily carry off pieces of DNA from other bacteria. (DNA is the code that contains the genetic instructions for all living organisms.) This trait enables bacteria to trade information on how to resist the antibiotics we aim at them. It also makes them a cooperative work force for producing human insulin. In a culture dish the size of your hand, billions of bacteria happily, if unconsciously, do the bidding of the molecular biologist.

Insulin, a hormone produced by small groups of cells in the pancreas called the islets of Langerhans, controls a person's blood sugar levels. When a person has diabetes, his or her pancreas produces either no insulin at all or not enough insulin. Insufficient insulin causes the sugar called glucose to build up in the blood and then be secreted in the urine. Symptoms of diabetes include fatigue, weight loss, muscular weakness, excessive thirst, and frequent urination. Untreated, it can lead to blindness, nerve damage, and kidney failure. There are two types of diabetes. The less severe kind, which typically appears in young adulthood, can usually be controlled by diet. The other kind, which typically appears during childhood or adolescence, requires close monitoring of blood glucose levels and daily injections of insulin. The latter kind accounts for about 10 to 15 percent of all cases of diabetes; people with this kind of diabetes are said to be "insulin-dependent."

Early diabetes treatment focused first on insulin harvested from human blood, and then on insulin harvested from pig's blood. Both methods depend on limited sources and require extreme care to keep the injected insulin disease-free. As a source of insulin, bacteria are superior to humans and pigs on both counts. They also take up less space and are cheaper to feed than humans or pigs.

To set their bacteria factory in motion, scientists remove from human DNA the section that controls insulin production. They then insert this bit of human DNA into the bacteria's DNA. Although bacteria have no use for insulin themselves, they industriously produce insulin when their new bit of DNA tells them to. As the bacteria divide to form new bacteria, the new bacteria, too, carry the DNA that tells them to produce insulin. To keep the factory going, the scientists simply provide food to keep the bacteria culture growing and remove older parts of the culture to isolate and purify the insulin. Because the DNA added to the bacteria originally came from humans, the insulin produced by the bacteria is identical to human insulin. Basically, it *is* human insulin—produced in a bacteria factory.

Does the ability of bacteria to "trade" DNA with other bacteria make it possible for them to develop immunities to our antibiotics? (See page 125: The Growing Resistance of Bacteria to Our Medicines)

Could it be possible to cure diabetes by inserting new genes that regulate insulin production? (See page 178: Genetic Engineering)

MONKEY SEE, MONKEY BE: XENOTRANSPLANTS

The replacement of damaged or diseased organs sounds as sensible as replacing the worn tires on your automobile. As with an automobile, you need to find the right model, size, or brand. Unlike an automobile, however, the body marshals powerful forces against what it perceives as a "foreign invader"—and anything from someone other than an identical twin is "foreign." Until recently, the emphasis in organ transplant technology was to suppress the body's rejection of the new organ with powerful drugs. When the most effective of these drugs, Cyclosporin A, was introduced in the 1980s, it made human organ transplants relatively routine. The main problem with drugs that suppress organ rejection is that they also suppress the recipient's immune system.

The simplest transplants in terms of availability are tissues, such as blood and bone marrow, that the donor regenerates. Next are paired organs, such as kidneys, which are necessary for survival. However, with absolutely vital single organs, such as the heart and liver, the only possible human donors are cadavers. Each year as many as three thousand people in the United States alone die while waiting for organs.

To many people, an obvious solution to the problem is to use organs from other animals. This type of transplant is called a xeno-transplant. *Xeno* means "foreigner," as in *xenophobia*, "a fear of foreigners." So far, because xenotransplants are still in the early experimental stages, they have been tried only as a last resort on patients who otherwise are given virtually no chance of survival. In 1984, an infant called Baby Fae was given a baboon heart at Loma Linda University Medical Center in California. She survived for twenty days. In 1992, doctors at the University of Pittsburgh trans-planted the first baboon liver into a thirty-five-year-old man, who survived for seventy days before dying from an infection.

Some doctors think that the most practical use of xenotrans-plants is as a temporary "bridge," to hold a patient over until a human organ becomes available. Another technique does not involve actual transplantation of the animal organ. In 1992, for example, doctors at Johns Hopkins University attached a pig liver to a twenty-five-year-old woman for four hours to filter her blood while she waited for a human liver.

For a xenotransplant to succeed, not only must rejection be suppressed, but the "foreigner" must be made less foreign. Dr. Thomas E. Starzl, one of the University of Pittsburgh liver transplant pioneers, developed the chimerism theory, according to which long-term acceptance of the baboon liver depends on cell migration—from the baboon liver to other parts of the patient's body, as well as from the body to the new liver. This could be likened to immigrants becoming assimilated into their new neighborhood.

Another way to make the foreign less foreign is to make it more human *before* the surgery. The latest trend in xenotransplant research is to genetically alter pigs to produce certain proteins produced by humans. Pigs are much more plentiful than nonhuman primates such as baboons. And because they are less like us, their use may raise fewer emotional and ethical problems with the public.

How can scientists genetically alter pigs? (See page 178: Genetic Engineering)

Are there any animals that can simply regenerate lost parts? (See page 140: A Call to Arms: Starfish Regeneration)

WHY WE DON'T LIVE FOREVER

We live and we die. Between the two, we age. Early in life, the processes of growth and aging overlap; then in early adulthood, the degenerative processes take over. Aging, which is different from the diseases that the elderly often succumb to, is a gradual decline in bodily structure and function. Eventually the body's diminished abilities to ward off disease result in death. The only multicellular animals that apparently don't die from the aging process itself are those that don't reach a maximum size at maturity and then stop growing; they continue to grow larger until disease or accident deals the final blow. These drinkers at the fountain of youth include tortoises and sharks.

The reason we age, and thus eventually die, is that we are programmed to. Cells reproduce by dividing, and each type of cell is able to undergo a particular number of doublings. Until the mid-1960s, scientists thought that theoretically our cells could live forever. But when human cells are grown in a laboratory, they divide a characteristic number of times and then die. Cells taken from elderly humans, which have already undergone a number of doublings, divide fewer times in the laboratory before dying than do cells taken from younger humans. Even when cells are transplanted from an old animal to a younger one of the same species, the older cells die at their predestined time. The only cells that do divide indefinitely are those of cancer.

The body's cells don't just do their thing for eighty or so years and then stop. As the body ages, the process of cell division itself slows down. There is an increase in the cells of what scientists call the "wear-and-tear" pigment. As much as we may try to hide them, we all show the outward signs of aging. Most noticeably, our skin becomes wrinkled and less elastic, and our hair grays and thins. As tissues atrophy, the body finds it harder to self-regulate such functions as temperature, blood pH and sugar levels, and gland secretions. This makes the body less resistant to infection. The sense organs become less sensitive. The person has trouble seeing and hearing—even tasting, for by old age, the number of taste buds on the tongue is reduced to about one-third of the original number. Cells that don't reproduce, such as brain cells, also lose function with age.

Although humans have a given life span, and the aging process is inevitable, it isn't always easy to distinguish the physical effects of aging itself from social and environmental effects. Boredom and lack of mental stimulation, for example, can diminish a person's mental functioning as surely as tired brain cells. And even a young person who sits in a chair all day will lose muscular strength. Though scientists are not yet able to lengthen our lives (they're working on it!), they can help more people to live out the normal life span and increase the number of years that are spent in relatively good health. Of course, scientific descriptions of aging leave out the fact that it's never too late to learn ballroom dancing or make new friends, that we're never too old to laugh, and that some of us really do become wiser with age.

Could scientists eventually replace the human genes responsible for aging with those of a shark or tortoise? (See page 178: Genetic Engineering)

We know that all individual humans will eventually die, but will humanity itself die out? (See page 113: The Future of Human Evolution)

MEMORY

Despite the efforts, and advances, of neurobiologists and psychologists, memory remains largely a mystery. Somehow the sensory input of a fleeting moment is stored in a way that allows us to recall it months, years, even decades later. We can retrieve the image of our first-grade teacher, or of a house we lived in many years ago. Where is that image when we are not remembering it? Does it have an actual physical existence? It is not surprising that philosophers, as well as scientists, are fascinated by questions about memory and the brain.

Scientists divide memories into three main categories: sensory memory, short-term memory, and long-term memory. Sensory memory extends just briefly beyond the moment an experience takes place. For example, a person given a momentary glimpse of a photograph of twenty people will forget most of the photograph almost immediately afterwards. Some researchers believe that one function of sensory memory is to give a person a brief period of time to decide what experiences should be transferred into short-term memory. Short-term memory lasts for a short while, perhaps minutes. It allows us to remember a phone number long enough to dial it, or the spelling of an unusual name long enough to write it down. Information can be organized to make it more easily held in short-term memory. One method is to organize information into so-called chunks. Short-term memory can hold from five to seven chunks at a time, but there is no limit to the size of the chunks. The more meaningful the chunks are, the longer they can be and the easier it is to retain them. Chunks of random letters are hard to remember for even a period of seconds. Words are much easier to remember, and words strung into a sentence are even easier to remember.

Our long-term memory contains huge stores of information of many types, from addresses and phone numbers to memories of highly emotional experiences. Many researchers think that long-term memories are organized in a manner similar to a file system. To recall a memory, one needs only to find the right category, or file heading. For example, to recall what you had for dinner when you celebrated your birthday in a fancy restaurant, you might go to the heading "restaurant," or "fancy restaurants." If you always order seafood when you eat out, you might go to the heading "fish." People also find it easier to recall a memory if they have just recalled a similar memory. Many people have experienced conversations with old friends where one memory of, say, a cooking fiasco brings up memories of other cooking fiascos, each more hilarious than the last. Recreating the mood or situation in which something was learned also enhances recall, regardless of whether the original situation was positive or negative.

Just where are long-term memories filed? The simplistic answer is in the brain, but scientists don't yet understand the physical basis for memory. Long-term memories probably cause some kind of permanent structural or chemical change in the brain. The majority of scientists base their ideas on a model of the synaptic activities of the brain. The synapses are the tiny gaps between neurons, or nerve cells. When one neuron transmits an impulse to another, the impulse leaps across the synapse between them. A single neuron may have a number of synapses, creating many potential connections. Researchers think that when an experience or learning takes place, it creates a particular pathway along synapses. The more a memory is recalled or rehearsed—for example, by dialing a phone number over and over—the more that particular path is strengthened. It is also possible that chemical changes take place in the neurons themselves.

We experience ourselves in relation to the world; each day is a constant stream of perceptions, thoughts, and emotional responses. Many of each day's experiences make it into our long-term memory.

Without our long-term memories, we would have no sense of identity; each day, we would struggle to navigate the world anew.

> *How do the senses transmit information to the brain?* (See page 151: The Senses)
>
> *What is the chemical basis for the brain functions that underlie memory?* (See page 59: The Chemistry of the Brain)
>
> *Do other animals exhibit memory?* (See page 144: Octopus IQ)

LANGUAGE ACQUISITION

Tourists are often impressed by how smart children in other countries are—they all speak foreign languages! Linguists and psychologists—and parents—have long been fascinated by how children acquire language. The linguists and psychologists maintain a running argument that can be summarized as "nature or nurture." Those who argue for nature claim that our brain is wired to learn language, especially grammatical structure. Those in the nurture camp claim that children acquire language solely through learning, in particular, through imitation. Those in the middle claim that the process of language acquisition is a combination of nature and nurture.

The most famous proponent of the nature school is the linguist Noam Chomsky, who based his ideas on the ease with which children of all cultures learn grammar—not necessarily the formal rules of grammar as they're taught in school, but the practice of grammar in everyday speech. Chomsky believes so strongly that language capability is unique to the human brain that he dismisses any apparent language capabilities of chimps as being by definition nonlanguage. (Researchers named one chimp Nim Chimpsky.) Some scientists speak of genes for language. A Canadian linguist found a family with a genetic defect that gave them problems with grammar. But even if a single gene could disrupt the normal ease with grammar, it wouldn't necessarily follow that a single gene is responsible for the complexity of grammar usage.

Even within the nature camp, there is disagreement. Some researchers believe that when the early human brain reached a certain size, humans began to use language. Others say that as humans gradually acquired language, their brains grew larger in response.

Some post-Chomsky researchers began to look at children's language development in tandem with their cognitive development—their perceptions of the world, and their growing need to communicate their perceptions and needs. When a child says, "Mommy sleepy," for example, it could mean that the child is sleepy or that the child thinks that Mommy is sleepy. Generally, children's language development lags behind their cognitive development. We've all seen children try in frustration to express themselves, sometimes ending up in tears in the process. In this sense, the child's needs to question and manipulate the world drive his or her language development.

Scientists speak of "critical periods," when a young child is best able to learn new skills such as language, walking, and sensorimotor coordination. Deaf children must learn sign language during the critical period if they are to be fluent in it, though children who lose their hearing later, between the ages of nine and fifteen, are able to learn sign fluently. Apparently this is because when they learned spoken language, the necessary neurological patterns for language were set. Though it would, of course, be completely unethical to experimentally deprive children of the opportunity to learn new skills, there have been a few tragic cases where children were kept in total isolation, usually in a small room, by abusive adults for long stretches of their childhood. Though these children eventually acquire some language skills, they never acquire a true ease with language structure.

After adolescence, the organization of the brain is basically complete, making language acquisition much more difficult. This is why it is much easier for children than for adults to learn a second language. Too early, however, can cause other problems. Recent research has shown that children, such as immigrants, who learn a new language before the age of eight sometimes forget their native language.

As early as two months of age, babies begin to use what researchers call "prelinguistic phrases"—babble with some of the structure and rhythm of speech. Even deaf babies babble. Researchers think that one purpose of the babble may be to encourage the pretend conversations that adults carry on with bab-bling babies, so the babies will form the necessary social bonds before they learn to speak. Certainly, many nonscientists have com-mented on how babies sound as if they're "really" talking. Perhaps some of those babies are commenting to each other on how adult speech is a precursor to the superior babbling of babies!

Can any animals acquire language? (See page 203: Chimpanzees and Language; and page 201: Dolphins and Language)

Can machines understand and use language? (See page 212: Artificial Intelligence)

DOLPHINS AND LANGUAGE

The dolphin and ape are two animals about which we feel compelled to ask: do they use language? We could study how these animals communicate without deciding whether or not to label it *language*. But by asking this question we are not studying just dolphins and apes; we are also studying humans. Scientists and philosophers want to know what, if anything, makes humans distinct from other animals. If it is language, then how does our language differ from other animal forms of communication? If our language doesn't set us off, then do other animals share our language abilities?

Dolphin and ape language studies differ mainly in emphasis. Researchers working with apes spend the majority of their time trying to teach the animals language and interacting with them. Those who study dolphins also train and interact with them, but they spend a lot of their time observing how dolphins communicate with each other. Of course, it's easier to spend your waking hours interacting with a chimp, which can join in many human activities, than it is with a dolphin.

Dolphins are undoubtedly intelligent. Their brains are as large as those of humans. However, the neocortex, with which one reasons and creates, is much thinner than in humans. Some scientists place the dolphin between the dog and chimp in intelligence; others between the chimp and human. There are even people— probably not scientists!—who claim that dolphins are more intelligent than humans.

Anyone who observes dolphins is struck by the large variety of sounds they make. Writers have described some of these sounds as whispers, squeaks, claps, clicks, moans, squawks, barks, rattles, chirps, and singing. Pulsed squeaks seem to signal distress; chuckling sounds seem to accompany caresses. One of the most important sounds is the signature whistle, whose frequency and amplitude make it distinctive to the individual dolphin. Some researchers have even suggested that dolphins call each other by name with these whistles.

Dolphins are great mimics, which can make their actual language capabilities harder to judge. They can mimic their trainers' words; however, birds such as parrots and mynah birds can also mimic many words, without any comprehension of them. Yet even if the dolphin doesn't understand the actual words, the mimicry could be an attempt to communicate with the trainer.

An important aspect of language ability, of course, is comprehension. One measure of comprehension is how well an animal can carry out a given command. Researchers in Hawaii taught dolphins the word *hoop* and then the word *through*. When the two words were combined, the dolphins responded to the command by going through the hoop. When they substituted the word *gate* for *hoop*, the dolphins understood the new command and went through the gate.

In one of the most famous dolphin communication studies, conducted in 1965, researchers placed two dolphins in separate tanks. The dolphins could hear but not see each other. The researchers taught one of the dolphins to push paddles for rewards. They gave the other dolphin similar paddles but no training. After much vocalization between the two tanks, the second dolphin had learned the paddle technique. Though this suggested that the first dolphin taught the second how to use the paddles, it is possible that the second dolphin used the sounds it heard to pick up information that the first dolphin hadn't intentionally communicated. Of course, even that would be a feat of intelligence, if not language.

Do dolphins use their vocal abilities for any other purposes? (See page 162: Dolphin Echolocation)

How do humans learn language? (See page 198: Language Acquisition)

CHIMPANZEES AND LANGUAGE

Can chimps use language? Well, they can express themselves using symbols in a particular order, and they can understand spoken English. To the average layperson that looks a lot like language. But the answer depends on how you define language, and therein rages a controversy. Closely linked to the definition of language, and at the philosophical heart of the argument, is the question of what, if anything, makes humans distinct. Those of the continuity school believe that humans are just at the end of a continuum—more developed cognitively, but not drastically different from our nearest animal relatives. Those of the discontinuity school believe that humans are unique, and that language is a trait with no close evolutionary precedent in any other animals. Thus, to those of the discontinuity way of thinking, chimps are by definition incapable of using true language.

A rough definition of language is the use of symbols within a grammatical structure, or syntax, to express one's perceptions of the world. By this definition, an animal that is trained to use symbols simply to request particular foods or toys is not using language; it is using mimicry to get a reward. Those who disparage ape-language research say that even when animals appear to go beyond such simple requests, there is an underlying gimmick. They cite the famous case of Clever Hans, a horse that appeared able to do arithmetic problems by stamping its hoof the correct number of times. It turned out that, unconsciously, his trainer was using body language to tell him when he reached the correct answer.

One of the most prominent ape-language researchers is a woman named Sue Savage-Rumbaugh, who came from a background in child development. Her star pupil is a pigmy chimp named Kanzi. Savage-Rumbaugh and her associates had been using a keyboard with lexigrams (symbols with no visual resemblance to the things they stand for) to teach language skills to Kanzi's adoptive mother, Matata. After two years Matata had learned only six symbols, so they decided to try teaching Kanzi. To their surprise, Kanzi already had picked up the symbols they were teaching Matata. His first morning at the keyboard, Kanzi pressed the keys for *apple* and *chase*, then picked up an apple, looked over at Savage-Rumbaugh, and ran off playfully. This was the beginning of Savage-Rumbaugh's new method of teaching animals language, based on the way children learn language—through constant, unstructured exposure to it.

After a year and a half, Kanzi had a vocabulary of about fifty symbols. At this point, the researchers became aware that Kanzi could also understand spoken language without the keyboard. When they were talking about lights, Kanzi would go over to the light switch and turn the lights on and off. This was particularly significant because the conversation was not addressed to him, nor did it concern him. Savage-Rumbaugh and her associates were also impressed that Kanzi could follow spoken commands that had two parts and involved an object out of view, such as, "Can you go to the colony room and get the telephone?"

Kanzi can clearly follow simple grammatical rules. In fact, he has made up a couple of his own. One involves using a keyboard symbol followed by a gesture; for example, pressing the symbol for *chase* and then pointing to the person he wishes to have chase him. When skeptics scoff at these rudimentary forms of grammar, Savage-Rumbaugh accuses them of holding a double standard. After all, Kanzi's language development is similar to that of a two-year-old human child. But whereas with a human child, they would view the same language usage as the rudimentary stages of true language development, with Kanzi they scoff. And to the argument that much of the time Kanzi uses language to request things such as food and toys—well, so does a two-year-old child. But he also uses language to tell what he has eaten, for example, or where he is going.

Perhaps Kanzi's feats of communication don't qualify as true language—perhaps language is solely the gift of humans. But the methods used in ape-language studies are finding fruitful application

in the teaching of communication skills to mentally disabled human
children.

> *If the genetic codes of humans and great apes are nearly
> identical, might apes share the human capacity for
> language?* (See page 173: DNA: Our Genetic Code)

> *Are there animals other than chimpanzees that might
> have the capacity for language learning?* (See page 201:
> Dolphins and Language)

MAPS

A map can be as simple as a few lines drawn in the sand or a sketch on paper napkin. Depending on our needs, the simple map may serve us as well as an elaborate map from a leather-bound atlas. If we're on a beach, a map in the sand showing the location of a refreshment stand is all we need. If we're in a restaurant, headed for a museum, a map on a napkin will do the trick. The basic information in such a map is, "This is here; that is there." Spatial relationships are the essence of all maps. Distances and sizes can be inaccurate, and overall shapes can be distorted, but as long as basic spatial relationships are correct, a map is valid. In other words, if you draw a line to represent the East Coast and put one dot for Boston, another below it for New York, and a dot below that for Washington, D.C., you've got a map. If you reverse the dots for Boston and New York, you've got a line and three dots, but no map.

Some psychologists and philosophers believe that our brain has an innate tendency to conceive of things in spatial terms and may even store certain types of memories in map form. We think in visual images, and visual images are inherently spatial. Even scientists, who describe the world in the most abstract terms, rely on visualization to express and manipulate their ideas. In many scientific fields, the map is the ideal way to describe phenomena. One of the beauties of maps, in science as well as in everyday life, is that the user can extract more information than the mapmaker needed to construct the map. In other words, only a limited amount of data is required to make a map, yet the skilled map reader can use it to answer a virtually limitless number of questions. For that reason, philosophers of science like to compare scientific theories to maps— a good theory often opens up avenues of inquiry beyond its original scope. Even our simple little map with the line and three dots could help a European plan an itinerary for the eastern portion of a trip to the United States.

Scientists map phenomena ranging from the subatomic to the cosmological, and every scale in between. The scientific map is far more than a record of discovery; it is an important tool of investigation. Computers play a large role in the modern scientific map. Atomic physicists "see" atomic particles by creating them in "atom smashers." These particles are very short-lived, and physicists infer their brief existence from tracks they leave in detectors. The detectors then create maps of the particles' charge, mass, and path. One of the most famous scientific maps, which scientists are still in the process of creating, is the Human Genome Project, the complete map of the DNA code that makes up the human chromosomes. Scientists use computer imaging to map all parts of the human body, including the heart and brain. On the global scale, scientists map weather patterns, erosion patterns, and the ozone layer. Geologists have long depended on mapping to study the chronology and relationships of the rock layers of Earth's crust. On the largest scale of all, maps help astronomers to explore our galaxy and beyond, out to the edges of the observable universe.

Maps enable scientists to move around in time as well as space. Paleontologists map the whereabouts and sequence of species that became extinct millions of years ago; climatologists use maps to project future climatic conditions. Eventually, geneticists will use what they learn from DNA mapping to manipulate the future, when they use gene splicing to alter a person's genetic map.

Maps describe the world for us; they also help us to locate ourselves in the world. Stores sell T-shirts printed with an image of the Milky Way and an arrow pointing to a spot, with the words "You are here." From subatomic particles to the universe, maps help us to understand and describe "here."

Can computers generate maps that we can "enter" to look around? (See page 214: Virtual Reality)

How are scientists "mapping" the human DNA code? (See page 176: The Human Genome Project)

DNA COMPUTING

You may not be able to read the evening news or a detective novel in a bowl of alphabet soup, but someday you may be able to find the solutions to complex mathematical problems in a bowl of DNA soup. In November 1994, a computer theorist at the University of Southern California at Los Angeles named Leonard Adleman startled the scientific world by announcing that he had constructed a simple computer using DNA. His work actually developed out of his interest in HIV, the virus that causes AIDS. Adleman modestly claims that it was only a matter of time before a computer theorist let loose in a molecular biology lab invented a DNA computer.

Adleman went easy on his first DNA computer and set it a relatively easy problem, one called the "traveling salesman problem." Given seven cities, a traveler must find a route that goes through each city exactly once. DNA, the nucleic acid that contains the genetic blueprint for all living organisms, contains a multitude of possible combinations of four subunits—adenine, thymine, guanine, and cytosine, referred to as A, T, G, and C. Using these letters as a code, Adleman assigned each city a code name consisting of a twenty-letter combination of these four letters. He then took the first ten letters of the code name for the first city of the salesman's journey and attached it to the first ten letters of the code name for the final city. The complementary sequences helped to hold the pieces together. When he dissolved his microscopic construction in water, the DNA strands joined together in longer sequences, or molecules. Adleman's job then was to determine which of the trillions of molecules formed contained the answer to his problem. Important clues were that the molecule had to begin with the name of the first city and end with the name of the last city, and had to be the correct length. It took Adleman a week to find the successful molecule.

A few months after Adleman's announcement, a Princeton University researcher named Richard Lipton and two of his students came up with a way for a DNA computer to crack the supposedly uncrackable codes of the U.S. National Security Agency. Though they have yet to actually make this much more complex computer, they believe it to be theoretically possible.

The beauty of a DNA computer would be the speed of its almost instantaneous chemical reactions. In other words, a DNA computer would have tremendous powers of what is called parallel computing—working on many tiny problems at the same time. Scientists believe a DNA computer could perform more calculations in a few days than all the computers ever built have performed up to now. DNA computers would also have incredibly large memories. A tank containing a pound of DNA in 1,000 quarts of liquid would have more memory than all of the computers ever built. The downside is that searching for the answer in a DNA computer would be time-consuming. Also, just as the letters in your bowl of alphabet soup get mushy when they stand around too long, after a while a DNA computer would dissolve.

Leonard Adleman used to think that a computer was a physical instrument that was "out there" in the world. Now he has begun to think that a computer is anything whose behavior we can interpret in computational terms. Oddly enough, DNA contains more codes than are used for life's genetic instructions. Perhaps there are other codes around, waiting to be discovered by computer scientists. Perhaps that bowl of alphabet soup will be of scientific use, after all.

Could DNA computers become "intelligent"? (See page 212: Artificial Intelligence)

PEAR: PRINCETON ENGINEERING ANOMALIES RESEARCH

Engineering anomalies? What are they—automobiles with wings? Helicopters with foam-rubber blades? Engineering anomalies is what Princeton University researcher Robert George Jahn calls the phenomena he studies. Others would use terms like *psychokinesis* or *paranormal*. The colloquial name for what he studies is *mind over matter*. Whose mind and what matter? The minds of his untrained volunteers. The matter in this case is computers specially designed to measure small but unambiguous changes in the performance of simple functions.

The story began in the later 1970s, when a graduate student at Princeton asked Jahn, then Dean of the School of Engineering and Applied Science, to supervise an independent study on psychic phenomena. Jahn, who wasn't particularly interested in the topic, was persuaded by the student's strong grades and her argument that the project would call upon her electrical engineering and computer skills. The student went on to design the machine that became the backbone of Jahn's work, a random event generator, or REG. The student eventually lost interest in the field of psychokinesis, but for Jahn it turned out to be the beginning of a whole new professional path. In 1979, he started the PEAR lab, to the dismay of many of his colleagues.

The REG is basically a computerized coin-flipper that electronically flips two hundred coins very rapidly and then counts the heads. In Jahn's experiments, volunteers mentally "will" the computer to favor either heads or tails. They don't even have to sit in front of the screen—volunteers on other continents have participated. Over the years, volunteers have been able to cause a small but statistically significant deviation from 50/50 that one would expect to occur by chance only 1 in 5,000 times.

Another PEAR machine is called Murphy, for Murphy's law. Murphy, also called the pinball machine, carries the technical moniker of "random mechanical cascade." Murphy takes 9,000 polystyrene balls to a height of 10 feet and then drops them, after which they fall through 330 pegs, to land in 19 bins. The entire process takes 12 minutes. Left on their own, more balls will fall in the center bins than the outer ones. The task of the volunteers is to will more balls to fall into bins either to the left or to the right of center. Here, too, volunteers have produced results that may be statistically significant—but only to the left. Of course, the important word in both random event generator and random mechanical cascade, of course, is *random*.

Though Jahn is the grandfather of such research, there are others in the field. In 1993, a Nevada researcher named Dean Radin started a privately funded lab at the University of Nevada called the Consciousness Research Laboratory. He has achieved random generator results similar to PEAR's.

Jahn and his associates meet with plenty of skepticism. Whatever the explanation for his results, the world could find plenty of practical applications for "mind over matter." Researchers are already working on ways for paralyzed people to communicate with computers via brain waves.

Does randomness increase or decrease on its own? (See page 21: Entropy)

Why do tossed coins tend to land 50 percent heads and 50 percent tails? (See page 75: Probability)

ARTIFICIAL INTELLIGENCE

We already have computers that can play a challenging game of chess or do complex mathematical calculations. But to artificial intelligence, or AI, researchers, a computer that could read the weekly supermarket tabloid and discuss the latest scandals would be far more remarkable. This is because AI studies how humans think and act in everyday life. The computer is the tool that AI workers use as they study how the mind works, and the computer program is the language they use to articulate their theories.

Most of our actions are much less methodical than chess games and math solutions. To function as a human mind does, a computer must be able to deal with the unexpected, make sense of incomplete data, decide which of a word's many meanings apply to a situation, recognize patterns, and learn from its mistakes. Basically, the computer must exhibit common sense. It must learn that the sentence "Aunt Irma hit the nail on the head" may mean that Aunt Irma was learning how to hang pictures, but is more likely to mean that she sized up a situation perfectly.

As long ago as 1950, a British mathematician and AI pioneer named Alan Turing proposed the Turing test for determining whether or not a machine was intelligent. Basically, a machine is intelligent if it can fool a person into thinking that he or she is conversing with another person. Since 1992, an annual competition has been held to award a prize to the computer program that fools the most judges. The judges hold conversations via computer terminals with both humans and with competing programs on such themes as "Pets" and "Bad Marriages." One tactic that seems to help a computer program fool judges is to evade specific questions with jokes.

A primary concept of AI is that of representation. How does the mind (or computer) represent the external world to itself? How does it manipulate those representations to understand and interact with the world, and to come up with new ideas? One challenge of AI is to give the computer an "awareness" of how it represents things and the ability to improve upon that process. Another crucial concept is that of knowledge. Knowledge is the entire store of information programmed into the computer. An intelligent computer must know how to structure a search of its knowledge to efficiently find a solution to a given problem. For example, if the computer is to choose what color to paint a baby's room, it might start by searching its knowledge about psychological studies on which colors humans find restful and which colors they find disturbing.

Subfields of AI with great potential practical application include speech recognition (e.g., computer secretaries that could answer their bosses' phones and tactfully turn down lunch invitations), robotics, and expert systems. An expert system is a store of knowledge about a specific subject, along with rules along the lines of "if this, then that." Researchers are working on the use of expert systems in medicine. For example, a family practitioner might use an expert system to augment his or her own knowledge in making a diagnosis. The physician could enter symptoms into the computer, which would then compare the symptoms to the hundreds or thousands of similar cases stored in its memory. Ideally, the expert physician would not use the expert system as a substitute for attentive, compassionate patient care, but as a useful tool.

Many AI workers are hugely optimistic that someday we will have computers that truly exhibit intelligent thought, common sense, and the ability to grow from experience.

How do humans "represent" the external world to themselves? (See page 151: The Senses)

Artificial intelligence requires the ability to store and access information. Is this similar to human memory function? How does human memory function? (See page 195: Memory)

VIRTUAL REALITY

Where cyberspace is both a technological and social phenomenon, virtual reality is basically a computer-based technology. In its ideal form, not yet achieved, it presents the user with the total illusion of a physical reality, a virtual world to inhabit and interact with. Even in its clumsier forms—heavy headgear and data glove, unconvincing graphics, limited repertoire of actions—its focus is the interaction between the user and the three-dimensional illusion. The potential applications of virtual reality are endless, and perhaps most exciting not in the gee-whiz realm of fantasy games, but where they can refine and extend the skills of people such as surgeons, engineers, chemists, and even psychotherapists.

Virtual reality depends on two types of apparatus: one to create the illusion of the virtual reality, and another to allow the user to interact with it. The user wears some kind of headgear that covers the eyes, creating a three-dimensional visual world, and sometimes headphones to create three-dimensional sound. At the same time, the user manipulates a handheld device called a data-gun to interact with the virtual world. A computer senses the user's hand and body movements and alters the virtual world accordingly. Turning one's head, for example, changes one's visual perspective in the virtual world.

Potential medical applications involve a technology called telepresence. The user interacts not with an illusory world, but with the image of a world that exists physically elsewhere. One aim of telepresence research is to allow surgeons to perform surgery in areas of the human body too small and/or delicate to work on by hand. Manipulating instruments attached to the hands, a surgeon would perform surgery on a three-dimensional projection of the patient. A computer tracking the surgeon's motions would send commands to robotic arms that would perform the actual surgery. Surgeons might also practice the night before on a virtual patient constructed to specification from diagnostic images taken of the real patient. The technology could also be used to perform surgery in places such as rural areas or battlefields, where no surgeon is available. The robotic arms would follow the motions of a surgeon performing the surgery on a telepresent image at a distant location. Some skeptics ask what happens if the instruments malfunction and there is no live surgeon to take over.

For architects and engineers, virtual reality allows one to test a design without building it. Because one can "walk around" in the virtual reality, an architect might try out the view from the top of a grand staircase, for example, or see how it feels to walk from the kitchen to the den—where the client's family will sit and play computer games.

Scientists such as chemists and biochemists can use virtual reality to model and study the structure and behavior of reality at the molecular and atomic levels. Here the value of virtual reality lies less in the illusion of reality—no sane scientist would expect to walk through a molecule—than in the ability to manipulate the three-dimensional images.

Psychotherapists have already used virtual reality to help treat patients with phobias such as extreme fear of heights. It is not too difficult to create the illusion of standing on a fifteenth-story balcony, or at the top of a ladder. Patients can gradually accustom themselves to the scary situation, knowing that they are in control and physically safe from harm.

Exercise-machine manufacturers have begun to design stationary bicycles equipped with virtual reality. The rider watches the changing view of country trails and feels the wind in his or her hair. The faster the rider pedals, the stronger the sensation of the wind. Perhaps an urban model should add the blaring of horns and touches of virtual olfactory reality, such as automobile fumes.

What is the relationship between virtual reality and cyberspace? (See page 216: Cyberspace)

If virtual reality is an electronic representation or simulation of reality, aren't our human senses also "virtual" to some degree? (See page 151: The Senses)

CYBERSPACE

Cyberspace is not spatial the way your living room or backyard is spatial. It is information—a huge network of electronic information open to anyone with access to a computer that is hooked into it. Spatially it has no shape: it is both local and global. Although some of its technologies, such as virtual reality, provide the user with a simulated physical experience, a word that describes cyberspace as a whole is *disembodied*. Instead of visiting a library, the citizen of cyberspace can retrieve information from countless institutions without leaving the house or office. Instead of visiting other people, the citizen of cyberspace can converse via computer with others who have similar interests. Seated at the computer, one can bank, shop, and make travel arrangements. And though travel still involves physical dislocation, many citizens of cyberspace go by the motto "Don't leave home without it"—meaning, of course, their computers.

Though computer capabilities grow dramatically year by year, what is newsworthy about cyberspace is not its technological advances, but the extent to which it influences how we think about and manage our daily lives. Take correspondence, for example. People used to write to each other. After mailing a letter, one would wait days or weeks for a response. Then people got used to using the telephone to make business arrangements and to keep in touch with friends and relatives, even those far away. Now many people depend on electronic mail, or e-mail. When you send an e-mail message, the message appears immediately in the electronic "mailbox." Like a telephone message, it is transmitted immediately, but like an old-fashioned letter, it is available to the recipient to read (or ignore) at any time.

Just as one can turn on the computer at any time to read one's e-mail, one can also participate in group discussions any time one wishes. Like geographical space, cyberspace is divided and subdivided into navigable segments. First, one joins what is called a network—either one of several commercial networks, or the global noncommercial network called the Internet. Once on the "Net," one selects groups and then subgroups, formed to discuss everything from parenting to vampires to political reform.

In cyberspace, individuals can create their own "pages"—sites for which they design the text and graphics, limited only by their imagination and computer programs. If you want to buy an antique fountain pen, for example, you "visit" the sites of pen dealers. If you are interested in a particular pen, you click to view an enlarged image of it.

Not surprisingly, the same social and political issues that are important in face-to-face communication arise in cyberspace. These include trust, deception, how to deal with nuisances, and censorship. Because people in cyberspace don't see each other (unless they choose to), their relationships are often free of the usual constraints of age, class, gender, race, and such. Some people use this freedom to take on a different persona. While some feel this to be harmlessly liberating, others consider it a betrayal of other people's trust. Questions of how to deal with messages that are offensive, or controversial, raise the same strong feelings they do in real life (IRL, as it is called). Because cyberspace is a relatively recent phenomenon, people—and governments—have to work out how regulations that govern speech and print relate to communication in cyberspace.

Cyberspace is here to stay. Socially, it is important that it be available to all segments of society. And individually, it's probably good to remember that its purpose is to broaden, not be a substitute for, activities IRL.

What is virtual reality, and how does it fit into cyberspace? (See page 214: Virtual Reality)

Are we likely to someday encounter artificial intelligences as "citizens" of cyberspace? (See page 212: Artificial Intelligence)

Could some ambitious person (or AI) make a map of cyberspace, if it isn't a physical place? (See page 206: Maps)

Index